EAI/Springer Innovations in Communication and Computing

Series editor
Imrich Chlamtac, European Alliance for Innovation, Gent, Belgium

Editor's Note

The impact of information technologies is creating a new world yet not fully understood. The extent and speed of economic, life style and social changes already perceived in everyday life is hard to estimate without understanding the technological driving forces behind it. This series presents contributed volumes featuring the latest research and development in the various information engineering technologies that play a key role in this process.

The range of topics, focusing primarily on communications and computing engineering include, but are not limited to, wireless networks; mobile communication; design and learning; gaming; interaction; e-health and pervasive healthcare; energy management; smart grids; internet of things; cognitive radio networks; computation; cloud computing; ubiquitous connectivity, and in mode general smart living, smart cities, Internet of Things and more. The series publishes a combination of expanded papers selected from hosted and sponsored European Alliance for Innovation (EAI) conferences that present cutting edge, global research as well as provide new perspectives on traditional related engineering fields. This content, complemented with open calls for contribution of book titles and individual chapters, together maintain Springer's and EAI's high standards of academic excellence. The audience for the books consists of researchers, industry professionals, advanced level students as well as practitioners in related fields of activity include information and communication specialists, security experts, economists, urban planners, doctors, and in general representatives in all those walks of life affected ad contributing to the information revolution.

About EAI

EAI is a grassroots member organization initiated through cooperation between businesses, public, private and government organizations to address the global challenges of Europe's future competitiveness and link the European Research community with its counterparts around the globe. EAI reaches out to hundreds of thousands of individual subscribers on all continents and collaborates with an institutional member base including Fortune 500 companies, government organizations, and educational institutions, provide a free research and innovation platform.

Through its open free membership model EAI promotes a new research and innovation culture based on collaboration, connectivity and recognition of excellence by community.

More information about this series at http://www.springer.com/series/15427

Angelo Beltran Jr. • Zeny Lontoc • Belinda Conde
Ronnie Serfa Juan • John Ryan Dizon
Editors

World Congress on Engineering and Technology; Innovation and its Sustainability 2018

Springer

EAI
RESEARCH MEETS INNOVATION

Editors
Angelo Beltran Jr.
Adamson University
Manila, Philippines

Zeny Lontoc
University of Perpetual Help System DALTA
Las Pinas, Philippines

Belinda Conde
University of Perpetual Help System
DALTA
Las Pinas, Philippines

Ronnie Serfa Juan
Cheongju University
Chungchongbukdo
Cheongju-si, Korea (Republic of)

John Ryan Dizon
Bataan Peninsula State University
Balanga, Philippines

ISSN 2522-8595 ISSN 2522-8609 (electronic)
EAI/Springer Innovations in Communication and Computing
ISBN 978-3-030-20906-3 ISBN 978-3-030-20904-9 (eBook)
https://doi.org/10.1007/978-3-030-20904-9

This Springer imprint is published by the registered company Springer Nature Switzerland AG.
The registered company address is: Gewerbestrasse 11, 6330 Cham, Switzerland

Foreword

We are pleased to present to you the proceedings of the first edition of the 2018 European Alliance for Innovation (EAI) International Conference on World Congress on Engineering and Technology; Innovation and Its Sustainability (WCETIS). This conference has become an instrument to gathered researchers from the academe and different field of discipline in engineering internationally, contributing to the enhancement of society through engineering and ICT.

The technical program of WCETIS 2018 consisted of 12 full papers. The conference has only one main track divided into seven general topics as follows: (1) Industrial Engineering and Healthcare; (2) Advanced production, processing, and manufacturing; (3) Sustainable Infrastructure; (4) Water Resources Planning and Management; (5) Heat transfer and fluids; (6) Electronics and Electrical Engineering; and (7) Internet of Things. Together with the high-quality technical paper presentations, the technical program also featured one keynote speech, with two technical talks. The keynote speaker is Dr. Angelo Beltran Jr. from Adamson University, Philippines. The two invited talks were from Prof. Lorena Ilagan and Mr. Rolando Pula of University of Perpetual Help System DALTA, Philippines. Lorena Ilagan discussed Internet of Things (IoT), while Mr. Rolando Pula talked about LiDAR Technology for Resources Mapping. The purpose of the talk is to spread some technological advances happening in the country through the application of latest technology in addressing major challenges.

Regular coordination with the steering chair, Imrich Chlamtac – Bruno Kessler Professor, University of Trento, Italy through the EAI Conference Managers – Radka Pincakova, and Karolina Marcinova was essential for the success of the conference. We sincerely appreciate their support and guidance throughout the process of handling and hosting a conference. This conference will not also be possible without the help of very supportive and dedicated organizing committee team. Particularly, the following individuals: Ronnie Concepcion II – Local Chair, Rolando Pula – Technical Program Committee Co-chair; Sheily Mendoza – Sponsorships and Exhibits Chair, and Cyd Laurence Santos – Web Chair. We also gave our appreciation for those reviewers who take time to review the papers. Additionally, we also want to thank the top management of the University of

Perpetual Help System DALTA in supporting us and, lastly, for all the authors who submitted their works to WCETIS 2018 conference.

We are firmly convinced that WCETIS conference provides a good venue for all different engineering fields in sharing their ideas, researchers, works, and scientific findings that promote advances in the seven major topics of the conference contributing to the betterment of mankind. We also expect that the future WCETIS conference will be much more successful and bigger in terms of number of high-quality papers in different engineering fields surpassing the previous volume.

Lorena Ilagan

Conference Organization

Steering Committee

Imrich Chlamtac	Bruno Kessler Professor, University of Trento, Italy

Organizing Committee

General Chair

Lorena Ilagan	Dean, College of Engineering – University of Perpetual Help DALTA Las Piñas City, Philippines

General Co-Chairs

Rolando Pula	Research Coordinator, College of Engineering – University of Perpetual Help DALTA Las Piñas City, Philippines

TPC Chair and Co-Chair

Rolando Pula	Research Coordinator, College of Engineering – University of Perpetual Help DALTA Las Piñas City, Philippines
Lawrence Charlemagne David	Research Coordinator, Maritime Academy of Asia and the Pacific

Sponsorship and Exhibit Chair

Sheily Mendoza	Chairman, Industrial Engineering Department – University of Perpetual Help System DALTA Las Piñas City, Philippines

Local Chair

Ronnie Concepcion II	Chairman, Electronics and Communication Engineering Department – University of Perpetual Help System DALTA Las Piñas City, Philippines

Workshops Chair
Ronnie Concepcion II — Chairman, Electronics and Communication Engineering Department – University of Perpetual Help System DALTA Las Piñas City, Philippines

Publicity & Social Media Chair
Mariz Vega — Faculty, Computer Engineering Department – University of Perpetual Help System DALTA Las Piñas City, Philippines

Publications Chair
Belinda Conde — Research Director, University of Perpetual Help System DALTA Las Piñas City, Philippines

Web Chair
Cyd Laurence Santos — Faculty, Computer Engineering Department – University of Perpetual Help System DALTA Las Piñas City, Philippines

Posters and PhD Track Chair
Raniel Suiza — Chairman, Mechanical Engineering Department – University of Perpetual Help System DALTA Las Piñas City, Philippines

Panels Chair
Edison Mojica — Chairman, Electrical Engineering Department – University of Perpetual Help System DALTA Las Piñas City, Philippines

Demos Chair
Cyd Laurence Santos — Faculty, Computer Engineering Department – University of Perpetual Help System DALTA Las Piñas City, Philippines

Tutorials Chairs
Shiella Marie Garcia — Chairman, Computer Engineering Department – University of Perpetual Help System DALTA Las Piñas City, Philippines

Technical Program Committee

Arnold Paglinawan	Professor, Mapua University, Philippines
Belinda Conde	Research Director, University of Perpetual Help System DALTA Las Piñas Campus, Philippines
Alfonso Loreto	School Director, University of Perpetual Help System DALTA Las Piñas Campus, Philippines
Ihsan Yassin	Professor, Universiti Teknologi Mara (UiTM), Malaysia
Dante Silva	Professor, Mapua University, Philippines

Contents

Part I
Industrial Engineering and Healthcare

A Web-Based "InstaSked" Appointment Scheduling System at Perpetual Help Medical Center Outpatient Department

Sheily Mendoza, Ranzel Cloie Padpad, Amira Jane Vael, Cindy Alcazar, and Rolando Pula

1 Introduction

Philippine health status indicators show that the country lags behind most of Southeast and North Asia in terms of health service outcomes. Since patient satisfaction is important, Philippine hospitals should take it up a notch by improving the current system. To achieve universal healthcare, the capacity of local government units to manage the local health system must be strengthened [1]. There are seven (7) types of waste in services. One of them is delay. Delay refers to customers waiting for service delivery. Currently, hospitals prefer traditional method of appointment – where patients are required to wait for a long time in the hospital before their consultation. The purpose of this research is to know the various causes of the long waiting time experienced by the patients and develop solutions through Industrial Engineering (IE) tools. Based on the study conducted by [2], the outpatient department is the main healthcare service for nonurgent patients, but long waiting time is a common problem for this particular department. Usually, the lack of an effective appointment system is the cause.

S. Mendoza (✉) · R. C. Padpad · A. J. Vael · C. Alcazar
Industrial Engineering Department, University of Perpetual Help System DALTA, Las Piñas City, Philippines

University of Perpetual Help System DALTA, Las Piñas City, Philippines
e-mail: shvmendoza@gmail.com; ranzel.padpad@yahoo.com; amirajane.vael@gmail.com; Imcindyyy12@yahoo.com

R. Pula
Graduate Studies, Mapua University, Manila, Philippines

School of Graduate Studies, Mapua University, Manila, Philippines
e-mail: pularolando@gmail.com

A. Beltran Jr. et al. (eds.), *World Congress on Engineering and Technology; Innovation and its Sustainability 2018*, EAI/Springer Innovations in Communication and Computing, https://doi.org/10.1007/978-3-030-20904-9_1

A web-based appointment system could possibly add to patient's satisfaction since it reduces total waiting time effectively compared to the traditional and usual queuing method [3].

The problems of OPDs of developing countries are long queues, inefficient hospital staffs, absence of staffs, etc. The utilization of time in the OPD setting can be studied further by timing the patient flow (TMS) [4]. The patient satisfaction correlates with the waiting time [5]. Thus, long waiting times have an adverse impact upon patients' perceptions of medical care.

OPD is described as the face of any hospital, as it is usually the first point where patients go in a hospital. Patient's impression is often influenced by OPD services [6]. Therefore, it is important that it provides excellent services for customer's satisfaction. It was stated in a study that increase waiting time at OPD causes negative impact on patient's satisfaction; hence, healthcare facility performance can be best defined by measuring the level of patient's satisfaction [7].

Evidently, Six Sigma is an approach for formulating a design and execution of new services, products, or processes [8]. It is an approach that is best for quality improvement that can be used to meet the expectations as well as the needs of the patients to improve profitability in healthcare organizations. In addition, a study has achieved significant reduction in waiting time in the outpatient services using the Six Sigma approach [9].

One of the several advancements that a hospital can use for managing different business-related process is the BPMS which will also be used by the proponents of this study. It is an integrated solution for business processes to meet the customer's needs better [10]. BPM in hospitals proves to be a beneficial methodology to improve both day-to-day and strategic operations [11].

2 Methodology

The research method used in this study was Six Sigma submethodology, DMADV (define, measure, analyze, design, and verify) for the first phase, while another tool, BPM (business process management), was used for the second phase. These two methodologies were used in defining the causes, gathering data, analyzing data, process redesigning, and process configuration, execution, and verification. Shown in Table 1 are the tools to be used in every phase of DMADV and BPM.

3 Results and Discussions

3.1 Define Phase

Determining the current system was the initial task done. This is to know the answers to the following questions: Is adjustment of the existing system necessary? What is the part of the system that needs improvement? Where should the adjustment start?

Table 1 Tools to be used in every phase of DMADV and BPM

Phase	Tools
Define	Process map, VOC (voice of customer), stockholders' analysis, surveys, interviews, CTQ tree
Measure	Arrival rate, waiting time, takt time
Analyze	Ishikawa diagram, data analysis, defects chart
Design/process design	Proposed process flow, diagrams
System configuration	Software requirements, policies, roles, prototype
Process enactment	Advantages vs. disadvantages table, takt time
Diagnosis/verify	Implementation, evaluation

Define phase is very important in conducting the DMADV and BPM Six Sigma submethodology, because this phase will be the guiding path to the direction of the study.

Flow Process Chart (Present System) Figure 1 shows the present system for consultation of patients in the OPD of Perpetual Help Medical Center.

Key Roles (Present System)

- Doctors
- Medical secretaries
- Patients
- Management

Slovin's formula was used in computing for a certain number of patients to be timed according to the population. Researchers takt timed three hundred ninety (390) patients for a month, so results could have a ninety-five percent (97%) confidence level and a margin of error of plus or minus three (±3).

Survey/Questionnaires This study used surveys as a way of obtaining voices, complaints, and suggestions from patients and medical secretaries. For the patient survey, the diversity sampling is used with one hundred (100) patients. For medical secretaries, expert sampling is used in which expert opinion is needed, with fifteen (15) secretaries chosen.

Time and Motion Study Slovin's formula was used in computing for a certain number of patients to be timed according to the population. Researchers takt timed three hundred ninety (390) patients for a month, so results could have a ninety-seven percent (97%) confidence level and a margin of error of plus or minus three (±3).

Critical-to-Quality Tree Based on the article by Team, Mind Tools Editorial, 2015, critical-to-quality (CTQ) trees help to translate broad customer needs into specific, actionable, measurable performance requirements that can then be used to deliver high-quality products and services. Researchers have gathered feedbacks from the management, secretaries, and patients. Through the gathered data, their needs and required performance from the system become known. Shown in Fig. 2 is the critical-to-quality tree analysis for the improvement of the existing system.

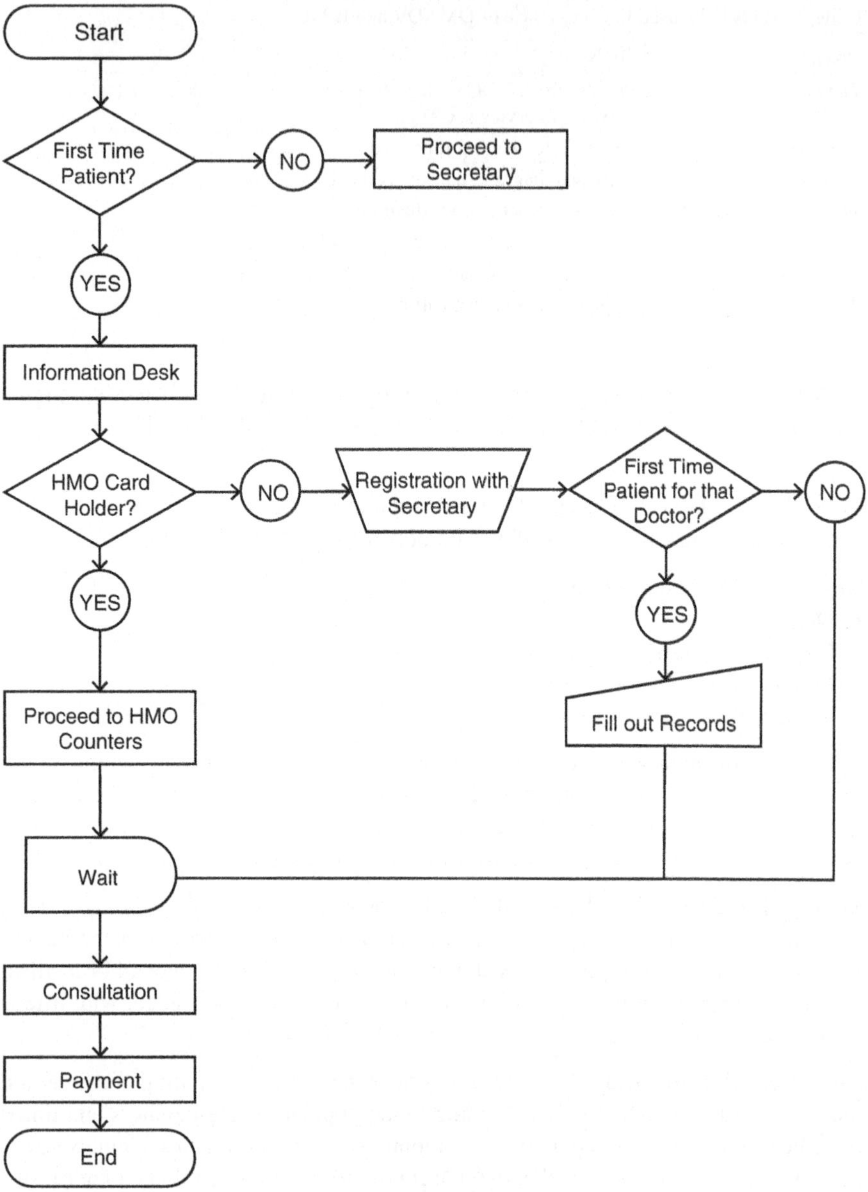

Fig. 1 Present appointment scheduling process for patients

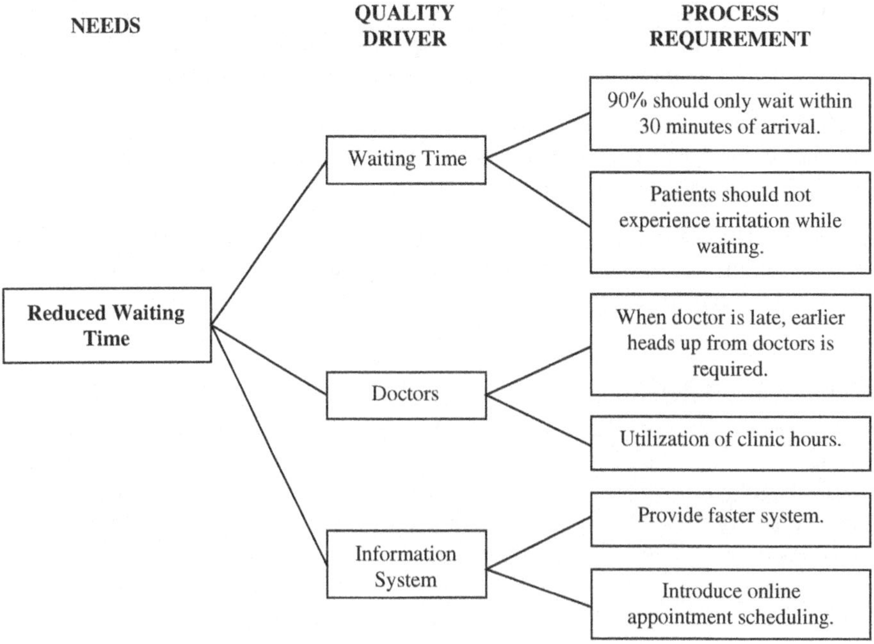

Fig. 2 CTQ tree

3.2 Measure Phase

According to Business Dictionary, time and motion study is the sequence of movements taken by the employee in performing steps with carefully observation to detect and eliminate redundant or wasteful motion, and precise time taken for each correct movement is measured. Researchers used TMS to measure the service time, waiting time, and arrival rate of patients in UPHDMC. It is a form of direct observation method. The starting point for TMS is when the patient writes his/her name in the schedule notebook, and it ends when he/she pays the bill. Results of the TMS in this study are shown below. All values are in minutes.

Time and Motion Study (Present System) Currently, there are six hundred (600) patients visiting outpatient departments every month. Researchers will takt time three hundred ninety (390) patients for a month, so results could have a ninety-seven percent (97%) confidence level and a margin of error of plus or minus three (±3).

As seen in Table 2, all specializations have a registration time of 0.2 minutes. The average waiting time of the six (6) specializations is eighty-five (85) minutes. A patient stays in the hospital on an average time of ninety-six (96) minutes.

Defects Chart Based on the current data, researchers have observed problems in the outpatient department and are shown below. Figure 3 shows the defects of the current process of the system inside UPHMC.

Ishikawa Diagram Ishikawa diagram in Fig. 4 shows the causes of customer dissatisfaction with regard to waiting time in percentage breakdown. The data is derived from the defects chart and surveys taken.

Table 2 TMS summary

Specialization	Registration time	Waiting time	Service time	Payment
Pediatrics	0.2	77.9	12.6	0.5
Internal medicine	0.2	76.3	11.5	0.6
General surgery	0.2	84.5	11.0	0.6
IM cardiology	0.2	81.9	8.8	0.6
IM pulmonary	0.2	96.9	11.4	0.6
Ortho surgery	0.2	89.7	10.0	0.6
Average	0.2	85	11	0.6

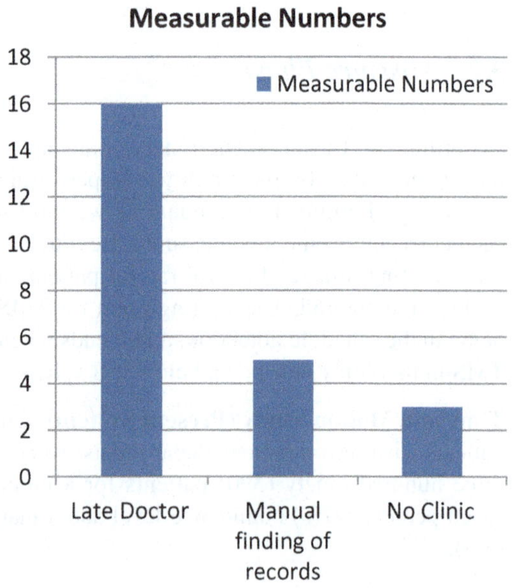

Fig. 3 Defects of the current process of the system inside UPHMC

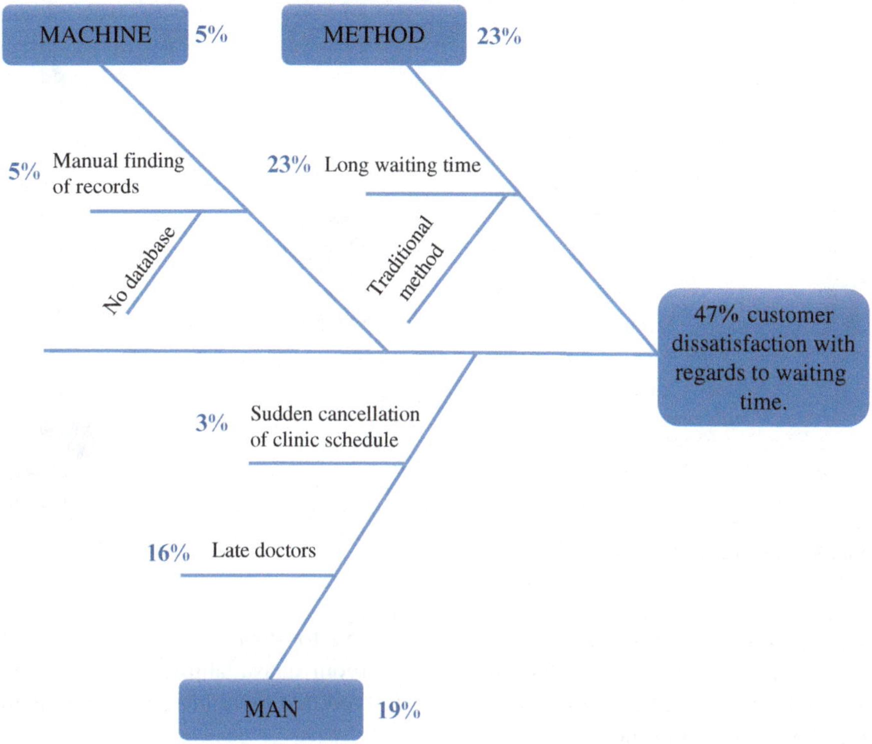

Fig. 4 Ishikawa diagram

3.3 *Process Design Phase*

In this phase, the proposed system design/workflow was established based on the data gathered from the above steps. Proposed workflow, process flow, and buffer rooms were created.

Proposed Workflow Diagram Figure 5 shows the flow of the scheduling proposed by the researchers. Patients should use computers or personal digital assistants connected to the Internet to schedule his/her appointment which will then proceed to the doctor, medical secretary, and management.

Proposed Process Flow Figure 6 shows the proposed appointment scheduling system for patients who want to avail the service of the doctor through online appointment.

Proposed Buffer Rooms Sixteen percent (16%) of the problem is contributed by the late doctors. It causes a longer waiting time for patients. Once the doctor arrives late, it will cause a domino effect on the patients listed. It could lead to problems such as not being able to consult the remaining patients due to room unavailability.

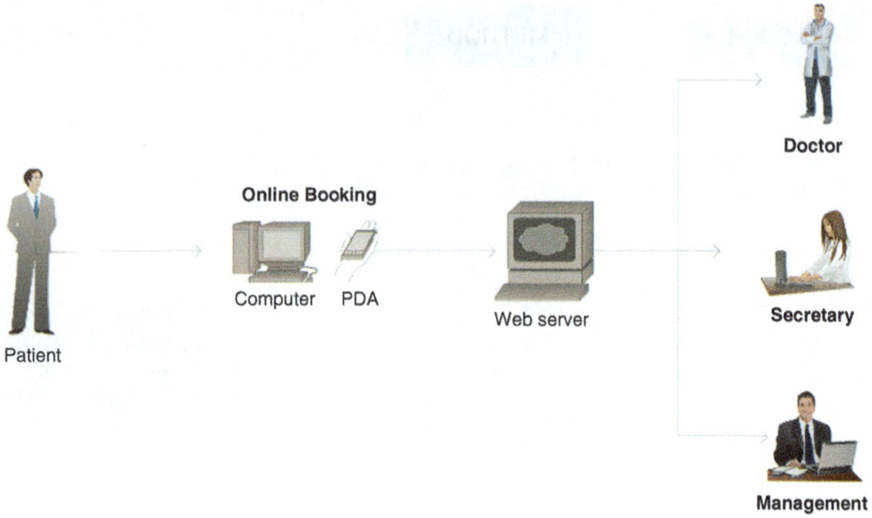

Fig. 5 Workflow diagram

Researchers proposed establishing buffer rooms for doctors that are in need of extending their clinic hours but can't due to room unavailability. Present room schedules were arranged, and proponents observed that there are vacant rooms in a certain amount of time.

3.4 System Configurator Phase

Software Requirements

- PHP
- Apache
- MySQL
- XAMPP

Prototype InstaSked is a web-based appointment scheduling system that allows patients to book their appointment in the comfort of their homes. It provides a real-time system wherein the users could monitor the current patient number accommodated. Text confirmation which includes all the appointment details is also sent to patients through this. It will be used by medical secretaries, doctors, and management. It enables secretaries to manage their doctor's patient list and send SMS whenever uncontrolled circumstances happen (e.g., late doctor, clinic cancellation) through logging in to their accounts. Doctors and the management can access the patient log listed according to schedule in the system by logging as well. Below is the screenshot of the InstaSked.

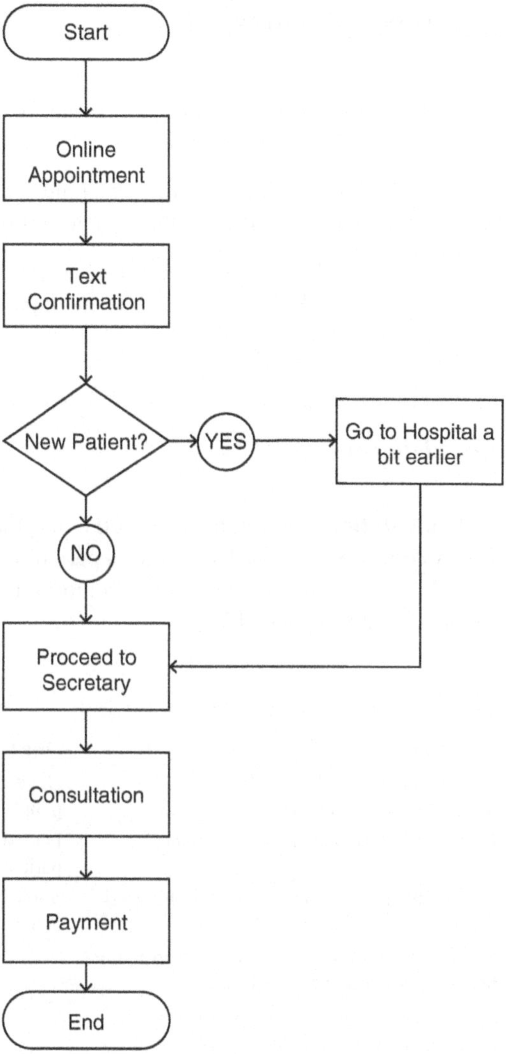

Fig. 6 Proposed process flow for patients

Policies of InstaSked

- The registration for booking an appointment is 2:1. For every two patients booked online, there will be one walk-in patient.
- If, for instance, the patient booked online is late, the next patient will be accommodated for the meantime regardless if booked online or walk-in.
- The secretary will inform the patients through sending a text message. There will be a buffer room available in case it will be occupied by a different schedule.
- Patients will have to show the secretary the confirmation text received upon booking the appointment.

3.5 Process Enactment

Advantages Versus Disadvantages (Table 3)

Time and Motion Study Through simulation, proponents used the time and motion study again to measure the changes brought by the proposed system.

The Table 4 above shows the time and motion study of the patients using the proposed system. With the proposed system, patients will spend only at least sixteen (16) minutes in the hospital.

3.6 Diagnosis/Verify

Implementation Researchers decided to use the parallel method. Both the old and new systems operate side by side for a period of time. Both are maintained and kept up to date so that in the event of the failure of the new system, the organization can fall back to the old one [12].

Table 3 Patients' view on the proposed system

Advantages	Disadvantages
Less hassle, no need to go directly to the hospital for appointment	Dependency in Internet connection (not all people have access to the Internet)
Easy and fast to book an appointment	Due to slow Internet connection, slot that the patients wanted might be taken by others
Allows the patients to choose their preferred schedule and doctor	Some patients have a lack of knowledge in using computers/Internet especially elders
Reduce the irritation of patients to wait in the hospital for a long period of time	
Patients receive notifications or updates	
Less contact with patients having contagious diseases	
Through this online appointment system, patients will increase their satisfaction of the hospital	
Availability of slots and doctors can be viewed	

Table 4 TMS summary

Elements	Ave. (in minutes)
Online appointment	1.118
Pre-consultation	3.125
Consultation	11
Payment	0.56

Table 5 Patients view on the proposed system

Advantages	Disadvantages
Less hassle, no need to go directly to the hospital for appointment	Dependency in Internet connection (not all people have access to the Internet)
Easy and fast to book an appointment	Due to slow Internet connection, slot that the patients wanted might be taken by others
Allows the patients to choose their preferred schedule and doctor	Some patients have a lack of knowledge in using computers/Internet especially elders
Reduce the irritation of patients to wait in the hospital for a long period of time	
Patients receive notifications or updates	
Less contact with patients having contagious diseases	
Through this online appointment system, patients will increase their satisfaction of the hospital	
Availability of slots and doctors can be viewed	

This method requires the traditional method of scheduling an appointment and InstaSked to run simultaneously. Once the new system is gradually accepted and widely used by the patients, the present system shall be eliminated.

Evaluation Testing and evaluation allows the client/customer to view the prototype and to give his/her views. Changes and improvements are agreed and further work carried out [13]. Focus groups were identified by the researchers for them to give their views and opinion. Proponents used a survey questionnaire in evaluating the prototype. The user groups are the patients, management, and medical secretaries. Table 5 shows the advantages and disadvantages from the patients' view.

4 Conclusion

Consultation patients of OPD in Perpetual Help Medical Center have been experiencing an average of eighty-five (85) minutes waiting time and forty-seven percent (47%) dissatisfaction with the present system of the hospital. These are due to patient irritability, traditional scheduling system, late doctors, and clinic cancellations.

With the proposed system, the average time patients spend in the hospital reduces from ninety-seven (97) minutes to sixteen (16) minutes. It decreased by eighty-three percent (83%) from the present system's time. It eliminated forty-two percent (42%) of the dissatisfaction experienced by the patients.

Acknowledgments The authors would like to thank all those who help in the completion of the project, especially UPHMC for allowing the study to be conducted in their vicinity. Special mention to the hospital staffs in supporting the study, providing all the necessary help the researcher needs.

Recommendations The researchers would like to recommend that a hospital management system should be applied. This would greatly help the doctors and secretaries in tracking their patients' record. A database that includes patient's records, results, consultations, and the like is highly recommended.

Creation of a mobile application version of InstaSked is also recommended, because it would be easier for the patients.

References

1. Romualdez, A. G., Rosa, J. F., Flavier, J. D., Quimbo, S. L., Hartigan-Go, K. Y., Lagrada, L. P., & David, L. C. (2011). The Philippines health system review. *Health Systems in Transition, 1*(2), 1–30.
2. Panaviwat, C., Lohasiriwat, H., & Tharmmaphornphilas, W. (2014). *Designing an appointment system for an outpatient department*. Bangkok: IOP Publishing. https://doi.org/10.1088/1757-899X/58/1/012010.
3. Nazia, S., & Ekta, S. (2014). Online Appointment Scheduling System for Hospitals–An Analytical Study. *International Journal of Innovations in Engineering and Technology (IJIET), 4*(1), 21–27.
4. Manna, N., Samsuzzaman, M., & Das, S. (2014, July). A time motion study in the OPD Clinic of a Rural Hospital of West Bengal. *Journal of Dental and Medical Sciences (IOSR-JDMS), 13*(7), 34–37.
5. Pillay, D. I. M. S., Ghazali, R. J. D. M., Manaf, N. H. A., Abdullah, A. H. A., Bakar, A. A., Salikin, F., Umapathy, M., Ali, R., Bidin, N., & Ismail, W. I. W. (2011). Hospital waiting time: the forgotten premise of healthcare service delivery? *International Journal of Health Care Quality Assurance, 24*(7), 506–522. https://doi.org/10.1108/09526861111160553.
6. Pandit, A., Varma, E., & Pa, A. (2016, 2016). Impact of OPD waiting time on patient satisfaction. *International Education and Research Journal, 2, 2*(8, 8). ISSN 2454-9916. Available at: http://ierj.in/journal/index.php/ierj/article/view/423. Date accessed: 01 Jul 2019.
7. Aswar Nandkeshav, R., Kale Kalpana, M., Rewatkar Mangesh, P., Jain, A., & Barure, B. (2014). Patients' waiting time and their satisfaction of health care services provided at Outpatient Department of Government Medical College, Nanded (Maharashtra, India). *International Journal of Contemporary Medicine, 2*(2), 72–76. https://doi.org/10.5958/2321-1032.2014.01031.6.
8. Bandyopadhyay, J. K., & Coopens, K. (2005). The use of six sigma in. *International Journal of Quality & Productivity Management, 5*, 1–12.
9. DMADV: An Approach for Developing New Initiatives. (2008). Innovation Insights (The Pennsylvania State University). Office of Planning and Assessment.
10. Scheithauer, G., & Wirtz, G. (2008). Applying business process management systems – a case study. Bamberg, Germany.
11. Breyfogle, F. (2013, September 18). Business process management system in hospitals. Retrieved from Smarter Solutions.
12. Grid Communication Toolkit. (2016) Retrieved from Webgate.
13. Ryan, V. (2013). Testing and evaluating prototype-why. http://www.technologystudent.com/despro_flsh/evalintegr1.html.

Modeling the Structural Characteristics of Porous Powder Materials with Application Models of Casual Two-Dimensional Packaging

Oleksandr Povstyanoy, Oleg Zabolotnyi, Roman Polinkevich, Dmytro Somov, and Olha Redko

1 Introduction

Creating new porous powder materials (PPM) with guaranteed properties has become possible with the help of traditional powder metallurgy technology, but it is necessary to forecast and control the parameters of their structure in the manufacturing process, which include: granulomere composition of the charge, the shape of particles, the density of the molded workpiece, the quality of the contacts, shaping, porosity, density, and volume distribution. However, the methods of powder metallurgy do not always ensure the homogeneity of properties within the materials and do not present an opportunity to obtain the structural characteristics of materials on a qualitative level. There is a need to increase the efficiency of traditional technologies and also an introduce non-waste production of widespread products, save energy, reduce labor costs, and control the parameters of the structure of powder materials in the process of their manufacture, possibly through forecasting using modern simulation tools [1–3].

The prediction of laws of the formation of structures and properties of materials depends firstly on the geometric factors of the powder particles. In addition, the analysis of modern technological processes of powder metallurgy shows that the presence of correlation links between the constituents, structure, and properties is provided by all operations of the technological process, where the initial stage is the filling of molds with a powder, which determines not only the size, shape, density, productivity, safety, and culture of work, but also affects a number of important properties of the finished product. Therefore, modeling experiments of forecasting

O. Povstyanoy · O. Zabolotnyi (✉) · R. Polinkevich · D. Somov · O. Redko
Lutsk National Technical University, Lutsk, Ukraine
e-mail: povstjanoj@ukr.net; volynasi@gmail.com; r.polinkevych@lntu.edu.ua; somovd@rambler.ru; redkooi@ukr.net

A. Beltran Jr. et al. (eds.), *World Congress on Engineering and Technology; Innovation and its Sustainability 2018*, EAI/Springer Innovations in Communication and Computing, https://doi.org/10.1007/978-3-030-20904-9_2

the dependence of material properties on the technological parameters of obtaining products using analytical, numerical, and numerical–analytical methods by three-dimensional modeling hold great importance in this arena.

2 Literature Review

The analysis of structurally heterogeneous materials was carried out by scientific groups under the guidance of R.M. Kadushnikov, I.G. Kamenin, Yu. N. Kryuchkov, V.D. Rud, M.B. Shtern, and V.D. Rud. The peculiarity of these works is the modeling of the structural characteristics of materials made up of particles in regular and irregular forms. The simulation results were checked on field experiments with manufactured carbido states, which proved to be highly effective in modeling. The obtained results qualitatively reproduce the process of filling the press form [4–7].

In the works of K.K. Kuzhidlovskiy, M.M. Lobur, O.M. Matviykova, and T.V. Semenov, the authors considered information methods for calculating the physical and chemical characteristics of powder materials [8–10]. The peculiarity of these works is that only the chemical composition or aggregate state of the initial material is analyzed, and the restoration of oxides, electrolysis of materials, and thermal dissociation of compounds was observed.

In foreign literary sources, much of the work on creating geometry is heavily centered on fiber modeling and the very boundary of geometry in composite materials [11, 12].

Modeling methods for the uncertainty of geometry influence the strength of components at the initial filling stage [13–15], but modeling of the random placement of powder at the filling stage in the hopper, taking into account the physical parameters of components, is completely unexplored.

There is a great need for the development of a methodology for modeling the calculation of a real fill in PPM form, which represents a more realistic level of heterogeneity and is the starting point for identifying the physical basis of powder behavior at the filling stage for most of the cases that are currently determined as empirical and characterized by real experiments. Ideally, the theoretical results obtained in this way change the engineering view of the properties of future PPM.

Therefore, the aim of our study is to develop a methodology for a calculation model of random pores in PPM at the stage of placing the material into the hopper, taking into account the physical fundamental components for a two-dimensional case.

3 New Method for the Calculation of Physical Parameters

For filtering the PPM, it is necessary to use powders with large particle sizes, while, at the same time, it is necessary to use powders of small particle sizes to achieve a high fineness of purification. These contradictions predetermine the need to find new technological techniques and methods that would allow the creation of PPM with such structures, which provide the most optimal combination of performance characteristics.

In addition, it should be noted that the practice of using new materials for the foundations of metal powders shows that implementation in full volume of their strength and performance characteristics requires a significant increase in the prediction accuracy of the physical and mechanical properties of materials and the development of new methods of modeling that include a comprehensive analysis of the processes of formation materials.

By analyzing the majority of domestic and foreign models of consolidation powders, the authors proposed a new method for the calculation of physical parameters, which formed the basis for the study of real packages (two-dimensional case).

Let area $G \in R^3$ be filled with small metal balls (Fig. 1).

It is necessary to establish the integral characteristics of the material to be received, such as the dielectric penetration. To solve this problem, we use the following approach [16]. Consider that, as a result of filling the area G, the particles appear as stationary, isotropic, random fields $\{\xi_{ij}(u),\ u \in R^3\}$, which describe the dielectric permittivity of the population.

If the electric potential of the region is applied as ϕ, then the potential U inside the area is the solution to the Dirichlet problem:

$$\begin{cases} \sum_{i,j=1}^{3} \xi_{i,j} \frac{\partial^2}{\partial x_i \partial x_j} U = 0 \\ U \Big/_{\partial G} = \varphi \end{cases}, \tag{1}$$

Since the coefficients ξ_{ij} have a complex nature, then a direct solution to the problem is impossible. Therefore, using the averaging method is proposed. Assume that $\forall i,j \in Z\ \xi_{ij}$ is ergodic and area G is large enough to manifest this property. That is expected, but instead of Eq. (1), note that:

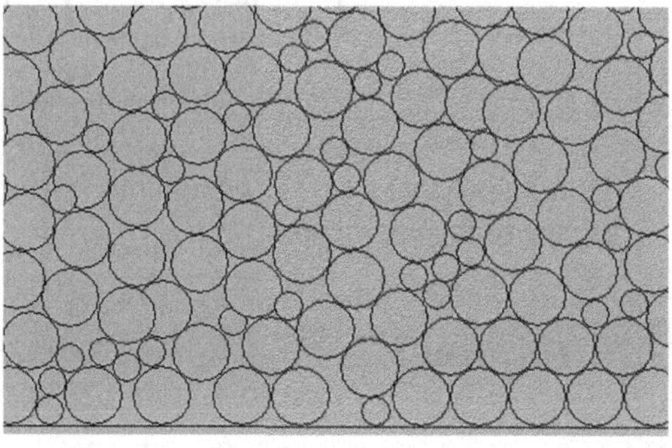

Fig. 1 Variant of filling a two-dimensional hopper with balls of appropriate size

$$\begin{cases} \sum_{i,j=1}^{3} \overline{\xi}_{i,j} \frac{\partial^2}{\partial x_i \partial x_j} U = 0 \\ U \Big/ _{\partial G} = \varphi \end{cases}, \tag{2}$$

with some "averaged" value of ξ_{ij}.

As part of this approach, the following simple cases were considered:

1. One-dimensional space, where $G \in R$. Instead of a random field $\xi(u)$, $u \in R$ considers a periodic function p: $R \to R$.
2. Multidimensional space, with the same periodic function p: $R^d \to R$, *де* $d > 2$.

3.1 One-Dimensional Case

Let $G = [-1; 1]$. In this case, the system in Eq. (1) can be rewritten as follows:

$$\begin{cases} x_\varepsilon''(u) = p_\varepsilon(u) x_\varepsilon(u) \\ x_\varepsilon(-1) = x_\varepsilon(1) = 1 \end{cases}, \tag{3}$$

where $p_\varepsilon(x) = p\left(\frac{x}{\varepsilon}\right)$and p: $R \to R$, which is a periodic function with a period of 1.

Decisions of this system are made through a probabilistic representation:

$$x(u) = M_u \exp\left\{-\int_0^\tau p_\varepsilon(\omega(s))\, ds\right\},$$

where $\{w(s), s \in [0, +\infty]\}$, known as the Wiener process, and $\tau = \inf\{s\colon w_u(s) \in \{-1; 1\}\}$, which is the moment of exit of the Wiener process from area G.

Since the measure of attending the Wiener process specified on R is absolutely continuous with respect to the Lebesgue measure, then the following equality holds:

$$\int_0^\tau p_\varepsilon(\omega(s))\, ds = \int_{-1}^{-1} p_\varepsilon(u) l_\tau(u) d,$$

where $l_\tau(u) = \int_0^\tau \delta_u(\omega(s))\, ds$, which is the local time of the Wiener process at point u.

If it is known that $l_\tau(u)$ is a continuous function, then for further transformations, we use the following theorem.

3.2 Theorem 1

For any continuous function f: $G \to R$ and a periodic function p: $R \to R$ with period 1, convergence is performed:

$$\int_{-1}^{1} p_\varepsilon(u) f(u)du \to \int_{0}^{1} p(u)du \int_{-1}^{1} f(y)dy, \varepsilon \to 0+$$

Proof. Let $\varepsilon = \frac{1}{n}$, $n \in N$ and replace the variables $x = un$:

$$\int_{-1}^{1} p_\varepsilon(u) f(u)du = \frac{1}{n}\int_{-n}^{n} p(x) f\left(\frac{x}{n}\right) dx = \frac{1}{n}\sum_{i=-n}^{n-1}\int_{i}^{i+1} p(x) f\left(\frac{x}{n}\right) dx$$

To prove the necessary convergence, make the following transformations:

$$\left|\frac{1}{n}\sum_{i=-n}^{n-1}\int_{i}^{i+1} p(x) f\left(\tfrac{x}{n}\right) dx - \left|\int_{0}^{1} p\right|(u)du \int_{-1}^{1} f(y)dy\right| = \left|\frac{1}{n}\sum_{i=-n}^{n-1}\int_{i}^{i+1} p(x) f\left(\tfrac{x}{n}\right) dx\right.$$
$$\left. - \frac{1}{n}\int_{-n}^{n} f\left(\tfrac{y}{n}\right) dy \int_{0}^{1} p(u)du\right| = \left|\frac{1}{n}\sum_{i=-n}^{n-1}\int_{i}^{i+1} p(x) f\left(\tfrac{x}{n}\right) dx - \frac{1}{n}p(x)dx\int_{i}^{i+1} f\left(\tfrac{y}{n}\right) dy\right|$$
$$\le \frac{1}{n}\left|\sum_{i=-n}^{n-1}\int_{i}^{i+1}\int_{i}^{i+1} p(x)\left(f\left(\tfrac{x}{n}\right) - f\left(\tfrac{y}{n}\right)\right) dydx\right|$$
$$\le \frac{1}{n}\sum_{i=-n}^{n-1}\left|\max_{|x-y|\le\frac{1}{n}}\left(f\left(\tfrac{x}{n}\right) - f\left(\tfrac{y}{n}\right)\right)\int_{i}^{i+1} p(x)dx\right|$$
$$= 2\int_{0}^{1}\left| p(x)dx \cdot \right|\max_{|x-y|\le\frac{1}{n}}\left(f\left(\tfrac{x}{n}\right) - f\left(\tfrac{y}{n}\right)\right)\Big|$$

Function f is continuous on the segment $[-1, 1]$; if it is uniformly continuous on the same segment, then the definition is definite:

$$\forall\varepsilon > 0\ \exists\delta_\varepsilon > 0 \forall x_1, x_2 \in [-1, 1]\ (|x_1 - x_2| < \delta_\varepsilon) \Rightarrow (|f(x_1) - f(x_2)| < \varepsilon)$$

Means $\forall\varepsilon > 0\ \exists N_0 \in N \forall n \le N_0$: $\left|\max_{|x-y|\le\frac{1}{n}}\left(f\left(\tfrac{x}{n}\right) - f\left(\tfrac{y}{n}\right)\right)\right| < \varepsilon$

Then, $\forall\varepsilon > 0\ \exists\ N_0 \in N\ \forall\ n \le N_0$:

$$\left|\int_{-1}^{1} p_\varepsilon(u) f(u)du - \int_{0}^{1} p(u)du\int_{-1}^{1} f(y)dy\right| \le 2\int_{0}^{1} |p(x)|\, dx \cdot \left|\max_{|x-y|\le\frac{1}{n}}\left(f\left(\frac{x}{n}\right) - f\left(\frac{y}{n}\right)\right|\right.$$
$$\left. < \varepsilon\int_{0}^{1} |p(u)|\, du\right)$$

That is, the necessary convergence is performed.

According to the proved theorem 1:

$$\int_0^\tau p_\varepsilon(\omega(s))\,ds \to \int_0^1 p(u)du \int_{-1}^1 l_\tau(y)dy,\ \varepsilon \to 0+$$

herewith:

$$\int_0^1 p(u)du \int_{-1}^1 l_\tau(y)dy = \int_0^1 p(u)du \int_0^\tau dy = \tau \int_0^1 p(u)du$$

and this means that the convergence of mathematical expectations is fulfilled:

$$M_u \exp\left\{-\int_0^\tau p_\varepsilon(\omega(s))\,ds\right\} \to M_u \exp\left\{-\tau \int_0^1 p(u)du\right\} \varepsilon \to 0+$$

That and the convergence of the solutions of the systems of equations give rise to the next theorem.

3.3 Theorem 2

Let x_ε be the solution of Eq. (3). Then, there is convergence:

$$x_\varepsilon \to x, \text{ at } \varepsilon \to 0+,$$

where x solves the problem:

$$\begin{cases} x''(u) = \int_0^1 p(s)ds\, x(u) \\ x(-1) = x(1) = 1 \end{cases} \tag{4}$$

which is a random field.

3.4 Theorem 3

Let $\{\xi_{ij}\}$ be a set of homogeneous isotropic random fields given on R^d $(d > 3)$. If the following conditions are executed:

1. $\forall i, j \in Z\ \exists\, A_1, A_2 \in R$, so that $\forall x \in R^d$, $\lambda_1, \lambda_2 \in R$: $A_1\|\lambda\|^2\ 6\ \xi_{ij}(x)\lambda_1\lambda_2\ 6\ A_2\|\lambda\|^2$

where $\|\lambda\|^2 = \sqrt{\lambda_1^2 + \lambda_2^2}$ and:

2. ξ_{ij} satisfies the degree condition of equally strong interchanges

then u_ε solves the task:

$$\begin{cases} \sum\limits_{i,j=1}^{3} \overline{\xi}_{i,j} \dfrac{\partial^2}{\partial x_i \partial x_j} u(x) = 0 \\ u \Big/_{\partial G} = g \end{cases}, \tag{5}$$

where G is a limited area with smooth boundary, g is some function, and u is the decision of the "average" task:

$$\begin{cases} \sum\limits_{i,j=1}^{3} \overline{\xi}_{i,j} \dfrac{\partial^2}{\partial x_i \partial x_j} U = 0 \\ U \Big/_{\partial G} = \varphi \end{cases}, \tag{6}$$

then:

$$M sup \{|u_\varepsilon(x) - u(x)| : x \in G\} = O\left(\varepsilon^\kappa\right), \quad \kappa > 0.$$

4 Results

Figure 2 shows an example of visualization fillings according to the developed computer simulation model [17].

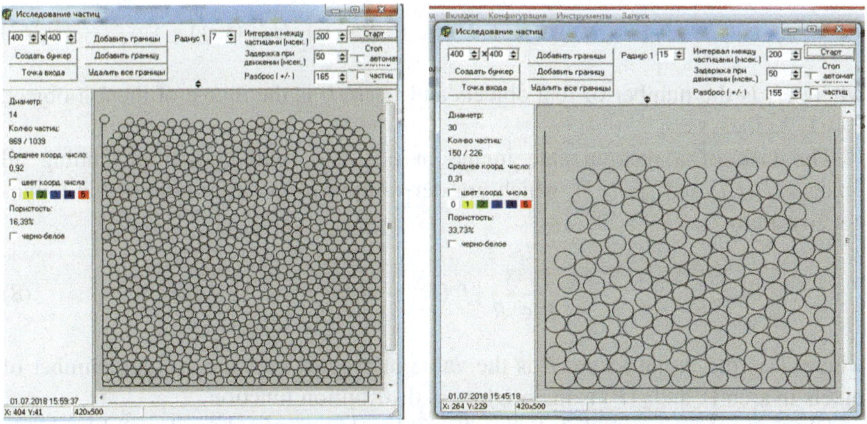

Fig. 2 Examples of visualization fillings of two-dimensional packaging in accordance with the developed computer simulation model

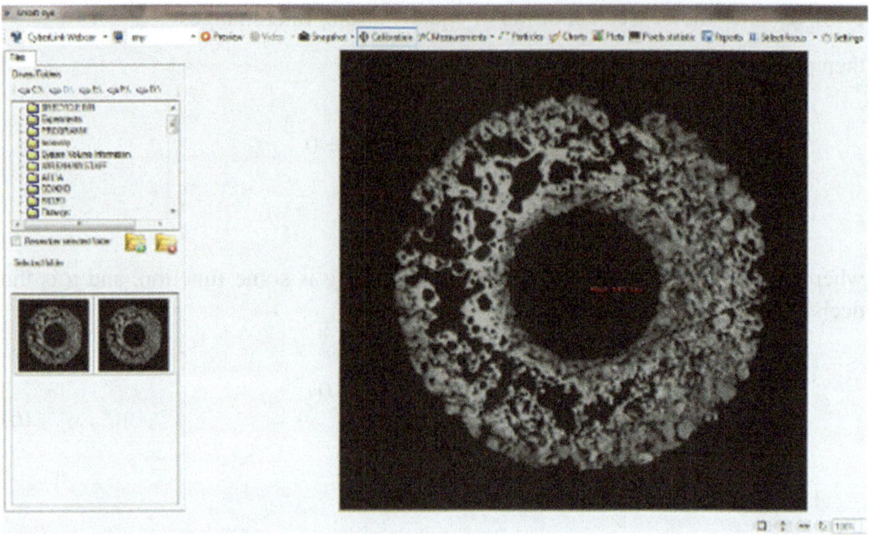

Fig. 3 Sample calibration process and real dimensional input with the Smart-eye® application

Analysis of the porosity distributions of PPM was carried out using the Smart-eye® application (Fig. 3). The software allows determining the necessary characteristics needed for a qualitative assessment of the structure of any material, including porous ones [18]. The algorithm of work is as follows: let $b_{i,j}$ be the outgoing image, where its value is equal to the brightness at point $i, j \in D$, where i=1, 2,... n and j=1, 2, 3,... m. Image $b_{i,j}$ is under real conditions and, also, a set of images of individual objects will be equal to:

$$b_{i,j} = H_1\,(i, j) + H_2\,(i, j) + \cdots + H_{s,k}\,(i, j) \tag{7}$$

where H_s is the number of real objects and H_k (i,j) is the image of the kth object, $k = 1, 2, 3, \ldots s$.

The task of a recognizable image in this case will consist of finding all objects $H_s(i,j)$ and H_k (i,j), which are determined from the criteria of the homogeneity area:

$$\frac{\max}{P \in R} |f\,(P) - m| \times T \tag{8}$$

where T is the initial value, P is the value in area R, m is the average number of pixels in area P, and $f(P)$ is the brightness distribution function.

Obtaining the porosity distribution (Figs. 4 and 5) gives a clear idea of the particle size distribution, provided that the radii (in fractions) are the same and characterize the differential distribution of the particles in the area of the cylindrical bunker [19].

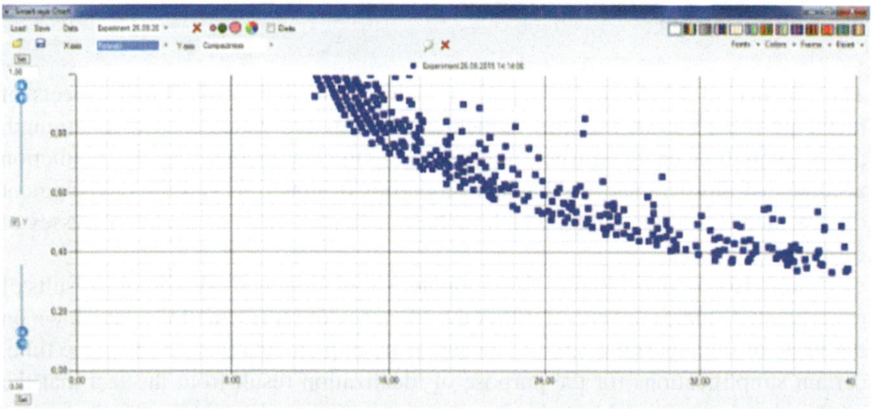

Fig. 4 Determination of the structural characteristics of the metal by a graphical method using the Smart-eye® application

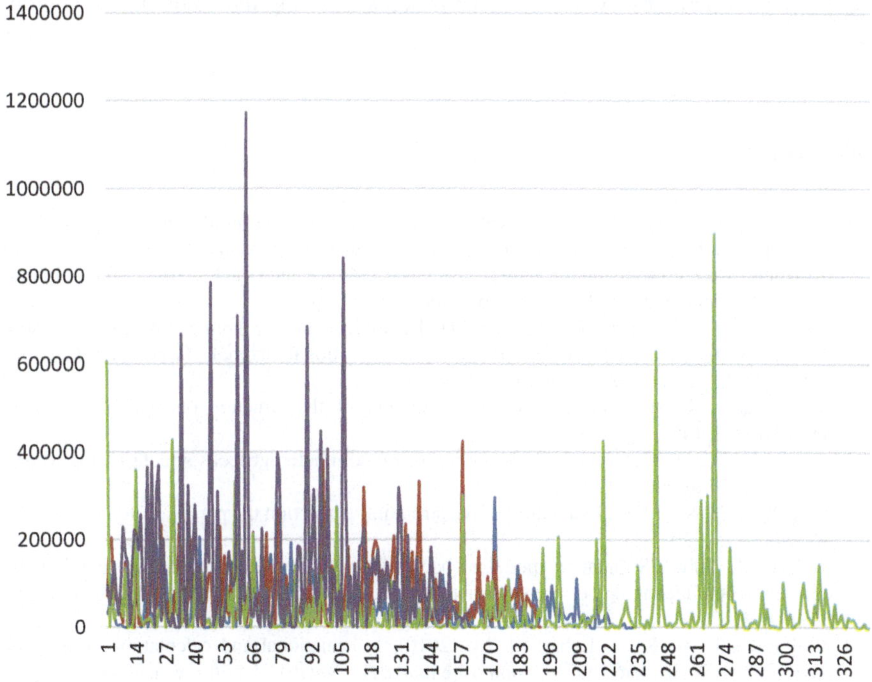

Fig. 5 Distribution of porosity change of porous powder materials (PPM)

5 Conclusions

The practice of producing new porous materials on the fundamental aspects of metal powders shows that the implementation in full volume of their strength and exploitation characteristics requires a significant increase in the prediction accuracy of physical and mechanical properties of materials and the development of new methods of modeling, which includes complex analysis of the processes of formation materials.

Comparison of data obtained by computer simulation methods with the results of experimental studies in general shows the adequacy of representations, laid down on the basis of the algorithm and the means of its implementation. At the same time, certain simplifications for the purpose of idealization result from the fact that the experimental curves are located below the corresponding data obtained by computer simulation methods. That is, the real tendency to compaction being higher than expected. It is possible to assume that, during the packing, there are processes of additional crushing, due to which there are additional conditions for the mutual movement of particles. As a result, the process of filling the voids becomes more significant.

References

1. Shibryaev, B. F. (1982). Porous permeable powder materials. *Powder Metallurgy, 2*, 140–151.
2. Boginsky, L., Reut, O., Piatsiushyk, Y. et al. (2001). The Development of Processes of Pressing of Articles from Powders on the Bases of Metals, Ceramics and Graphite. In: *15 International Plansee Seminar, 3*, pp. 197–209. Austria: Reutte.
3. Povstyanoy, A. Y., & Rud, V. D. (2014). Ecological and economic efficiency of using industrial production wastes for manufacturing constructional materials. *International Journal of Sustainable Development. Bulgaria, 19*, 89–94.
4. Kadushnikov, R. M. (1989). Geometric modeling of the structure of materials. *Powder Metallurgy, 2*, 140.
5. Kamenin, I. G. (1997). Simulation modeling of random heterogeneous structure of powders. *Powder Metallurgy, 7*, 302.
6. Kryuchkov, Y. N. (1996). Evaluation of the structural perfection the porous materials. *Powder Metallurgy, 5*, 220.
7. Rud, V. D. (2006). Simulation model of back filling of powder particles and using in the development of technology of preparation of a charge of carbide steels. *Modeling in Materials Science, 4*, 320.
8. Kuzhidlovskiy, K. K., & Lobur, M. M. (2010). Informational and calculation method for determining the characteristics of composite materials. *Powder Metallurgy, 6*, 23–30.
9. Matviykova, O. M., Shtern, M. B., & Mikhailov, O. V. (2002). Numerical modeling of pressing processes of powdered products of complex shape in rigid matrices: The influence of the pressing scheme on the density distribution. *Powder Metallurgy, 11*, 29–36.
10. Semenova, T. V., Shilko, S. V., & Kovtun, V. A. (2001). Mesomechanical analysis of granular materials under contact loading. *Mechanics of Composite Materials and Structures, 2*, 189–205.

11. Young, C., McMillan, A. J., Ravey, E., Verdicchio, J., Quinn, J., McIlhagger, A., & Buchanan, S. (2011). *The hybrid approach of a 3D textile composite finite element modelling technique at meso-scale level.* 18th ICCM, South Korea.
12. Potter, E., Pinho, S. T., Robinson, P., Iannucci, L., & McMillan, A. J. (2012). Mesh generation and geometrical modelling of 3D woven composites with variable tow cross-sections. *Computational Materials Science, 51*, 103–111.
13. Matthies, H. G. (2008). Stochastic finite elements: Computational approaches to stochastic partial differential equations. *Journal of Applied Mathematics and Mechanics, 88*(11), 849–873.
14. Moens, D., & Vandepitte, D. (2005). A survey of non-probabilistic uncertainty treatment in finite element analysis. *Computer Methods in Applied Mechanics and Engineering, 194*(12–16), 1527–1566.
15. Ainsworth, M., & Oden, J. T. (2000). *A posteriori error estimation in finite element analysis*. New York: Wiley Blackwell.
16. Povstyanoy, O. Y., & Dorohovtsev, A. A. (2018). Model studies of formation the powder filling in view of material properties on the basis of random packaging models (two-dimensional case). *Scientific Notes, 63*, 183–188.
17. Rud, V. D., Shuberko, V. V., & Povstyanoy, O. Y. (2014). Structural characteristics of blanks when filling molds with particles of irregular shape. *New Technologies and Materials, Automation of Production, 2*, 89–95.
18. Povstyanoy, O. Y., Suchuk, V. A., Macmillan, A., Rud, V. D., & Zabolotnyi, O. V. (2015). Metallographic analysis and processing of images of micro structure of nozzles for sandblasting, which are made by methods of powder metallurgy. *Powder Metallurgy, 3/4*, 234–240.
19. Povstyanoy Oleksandr (2018). *Theoretical and practical background of computer modelling implementation for solving of problems of powder materials forming*. The 2018 E-MRS Fall Meeting, Warsaw, Poland, September 17–20, 2018. Electronic data. Mode of access: https://www.european-mrs.com/multifunctional-advanced-composite-materials-idea-market-emrs#collapse70

Part II
Advanced Production, Processing and Manufacturing

Low-Cost, Automated, Rapid Bio Composting with SMS Monitoring System

Shiella Marie Garcia, Cyd Laurence B. Santos, Dexter C. Dolendo, Nicole Ann Roque, Arvin C. Marquez, and Rolando Pula

1 Introduction

A biodegradable material has the capability to decompose rapidly by the action of microorganisms. In waste management, these are named as biodegradable waste. Such wastes are table scraps, eggshells, leaves, and newspapers which are the usual wastes found in a household. Due to the lack of composting techniques, biodegradable waste ends up decomposing in landfills and produces methane which contributes as a main environmental threat [1].

Composting is the biological decomposition and stabilization of organic substrates under conditions that allow development of thermophilic temperatures as a result of biologically produced heat, to produce a final product that is stable, free of pathogens and plant seeds, and can be beneficially applied to land [2]. Stated in a program of San Mateo Country called RecycleWorks, composting reduces 30% of the waste streams. The composting process takes several months before the compost turns into usable soil. This is due to the effects of air, water, and thermal factors [3]. In the Philippines, composting is rarely engaged because of the time it takes to complete the process and limited space.

S. M. Garcia (✉) · C. L. B. Santos · D. C. Dolendo · N. A. Roque · A. C. Marquez
Computer Engineering Department, University of Perpetual Help System Dalta, Las Pinas, Philippines

College of Engineering, University of Perpetual Help System Dalta, Las Pinas, Philippines
e-mail: smpgarcia1972@gmail.com; cydlaurencesantos@gmail.com; dexter.dolendo@gmail.com; roque.nicoleann@gmail.com; marquez.arvin14@gmail.com

R. Pula
Graduate Studies, Mapua University, Manila, Philippines

School of Graduate Studies, Mapua University, Manila, Philippines
e-mail: pularolando@gmail.com

A. Beltran Jr. et al. (eds.), *World Congress on Engineering and Technology; Innovation and its Sustainability 2018*, EAI/Springer Innovations in Communication and Computing, https://doi.org/10.1007/978-3-030-20904-9_3

Records of DENR state that 8400–8600 tons of trash per day are produced by Metro Manila. Half of these are biodegradable waste, only one third of the biodegradable wastes are recycled into compost, and the rest are thrown in landfills [4].

There are several factors in the environment that affect the speed of the composting process. These factors include moisture and temperature. With the presence of proper moisture and temperature, the composting activity of the microbes quickens. Microorganisms can only make use of compost materials with proper moisture. During the process of composting, microorganisms produce heat which in turn increases pile temperatures [5].

Conventional monitoring systems require personnel at the scene to collect data to determine whether the temperature and humidity is within a limited range, resulting in waste of manpower and time, and inconvenience [6]. With the automation of the bio composting process, individuals can observe a rapid bio composting process with the help of proper monitoring of moisture and temperature value. Measuring soil moisture is an effective way to determine condition of soil and get information from it [7]. Shredding the biodegradable waste also helps fastening the breakdown process. As part of the monitoring system, a wireless transmission is needed to notify users about the composting process. SMS messages comprising preselected information are sent at predetermined intervals, such as at a certain time of day or when certain events occur, which is why the proponent chose SMS to notify users about the composting process [8]. The SMS notification also allows the individual to know if the composting process has been completed, making it easier for the individual to determine if the compost can be collected already.

2 Methodology

2.1 Theoretical Framework

Authorization required before the system starts to operate, by typing the password into prototype's User interface. The owner's mobile number can also be entered or changed by pressing the asterisk button and then entering the required password. Assigned number is where the SMS (notification regarding the completion of the process) will be sent. The SMS will send notification when the moisture level of the compost material reaches 18%. The System can also accept sensed temperature and moisture of the compost.

pH value is an important process parameter of the study, firstly because it is directly related to the decomposition achieved and the stability of the compost, and secondly because the optimal process temperature depends on pH value of the compost [9], and temperature rises due to resultant increase in molecular activity of substances on application of heat, which also connotes an increase in internal energy [10]. These variables are most important factors in monitoring the compost process.

Temperature and moisture value will serve as the input data in the process. These values will trigger and control the operation of the heater. For the initial process, the water pump applies a water and enzyme mixture to the compost material. A blower which acts as a heater turns on to increase the temperature of the compost material. The blower will stop when the temperature reaches 45 °C. Once the password is keyed in, the DC motor will turn on to drive the mixer which is used to promote aeration to the compost material by mixing it. When the temperature drops lower than 35 °C, the blower will turn on again, and the same process continues until the moisture value of the compost material reduces to 40% where the blower will cease its operation. The motor-driven mixer continues to operate until the moisture content of the compost material reaches 18%. At this stage, the process is already done.

Once the composting process is finished, the prototype will send an SMS notification to the user's mobile phone about the completion of the composting process. The prototype will also display the temperature and moisture value of the compost in the LCD screen interface.

2.2 *Conceptual Paradigm*

In the current process, biodegradable wastes are shredded down to tiny pieces, piled, and left for composting. These are then checked at random times to identify its status based on its appearance and texture. If the desired appearance and texture are reached, it will be assumed that the composting process is complete. This conventional method of composting takes 3 months to a year for the biodegradable waste to be fully composted.

As a result, conventional composting is a slow process due to improper monitoring and maintenance of temperature and moisture resulting to less biodegradable waste being recycled which contributes as one of the main environmental threats.

The prototype's input includes the owner's phone number, a password, and the moisture and temperature value of the compost. These parameter values can be seen on the LCD which changes with time and is used to trigger the heater operation during the process. The water pump initially applies the water and enzyme mixture to the compost material. The blower which serves as the heater works based on the changes in the parameter values. The motor-driven mixer operates to promote aeration to the compost material. The prototype then ceases its operation once the moisture value reaches 18%. Finally, the prototype sends an SMS notification regarding the completion of the process (Fig. 1).

2.3 *Description of the System*

First, the system will accept shredded biodegradable waste in its opening and into the vessel. Second, 240 ml of water and enzyme mixture will be applied to the compost material. Next, the motor-driven mixer will start mixing the compost material as the blower will then turn on. Constant monitoring of temperature and moisture is also applied in the process to promote microbial growth and faster decomposition. Finally, an SMS notification will be sent to the owner to notify that the composting process is complete (Fig. 2).

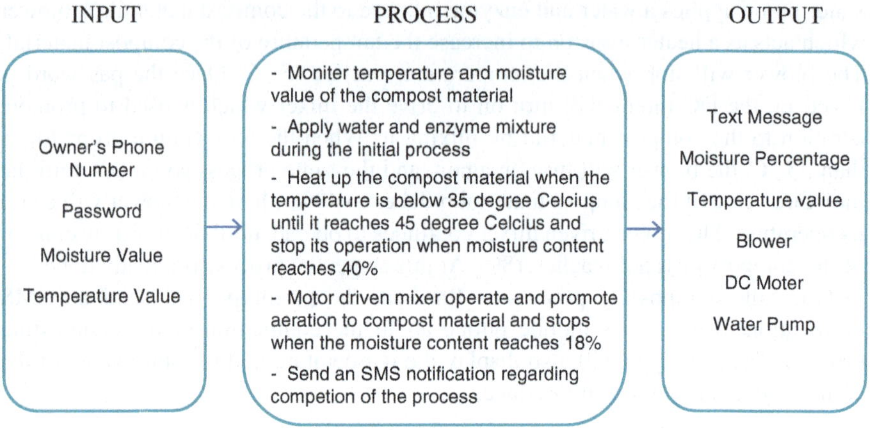

Fig. 1 Conceptual framework of the prototype

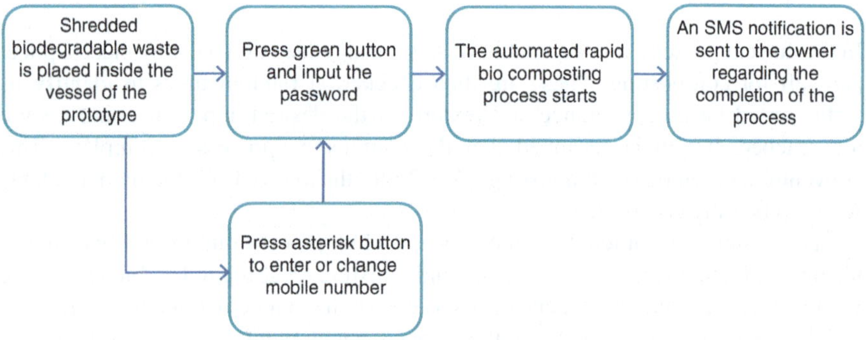

Fig. 2 Process of the proposed system

2.4 *Internal Design*

The 12v DC motor used to drive the mixer in the prototype could only mix up to 3 kg of shredded biodegradable materials. The system uses a blower instead of an actual heater to increase the temperature of the compost material inside the vessel until it reaches up to 45 °C. Because of that the temperature range is set only from 35 to 45 °C in order to avoid overheating of the blower.

2.5 *External Design*

Although the vessel of the prototype can hold up to 5 kg of compost material, the motor of the mixer could only drive up to 3 kg of the compost material, and because

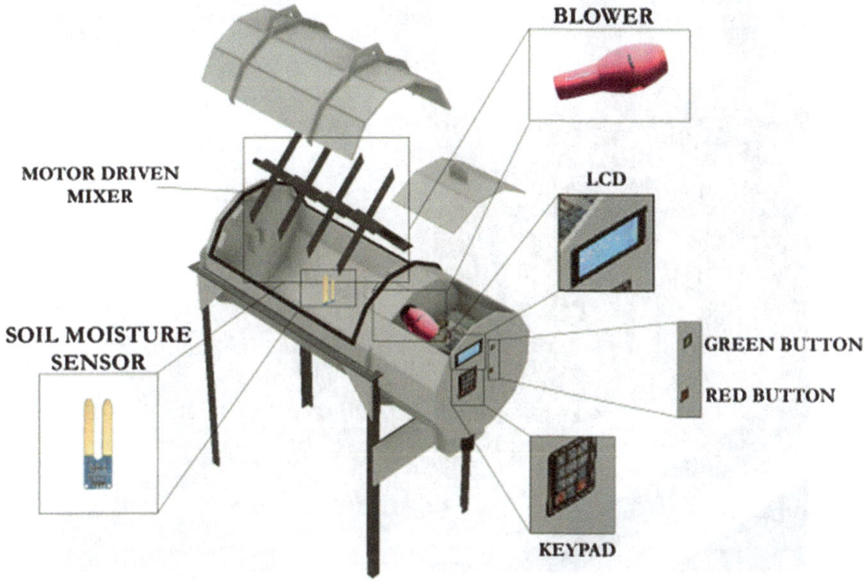

Fig. 3 Design of the prototype

of this, all of the testing that was conducted was only limited to 3 kg of the compost material. The motor-driven mixer is intended to be detachable from the vessel in order to collect the compost from the vessel (Fig. 3).

3 Results and Discussion

The prototype was tested using the materials from JM Purificacion Trading located at 23 Pasong Buaya II Imus, Cavite. Testing was performed to identify the types of biodegradable waste that can be accepted by the bio composting machine. The type of biodegradable waste will be identified and accepted, if and only if the motor-driven mixer will be able to operate and promote aeration for the material without any interruption in a span of 1 h.

Materials used as main ingredients were categorized as brown or green. Browns refer to materials such as dry rice husk, dry leaves, and branches that are all shredded. Browns (Shredded) refers to the shredded materials from JM Purificacion trading. Greens refer to materials such as kitchen waste, vegetable, and leaves that are shredded and containing moisture (Figs. 4, 5, 6, 7, 8, and 9).

The Table 1 shows the composting time of different compost materials. The browns alone completed the process in 22.25 h. For the 2:1 browns and greens ratio, the process was completed in 24.25 h. Finally, for the greens alone, the process was completed in 50 h. Thus, it can be observed that the compost material of browns

Fig. 4 Browns and greens mixed shown during the process of rapid bio composting

Fig. 5 Browns and greens mixed shown after rapid bio composting

Fig. 6 Greens alone shown during the process of rapid bio composting

Fig. 7 Greens alone shown after bio composting

Fig. 8 Shredded materials alone shown during the process of rapid bio composting

Fig. 9 Shredded materials alone shown after rapid bio composting

Table 1 Composting time of different compost materials

Component material	Quantity	Date/time started	Date/time completed	Number of h and min	Remarks
Brown set 1 (rice husk)	3 kilos	02-02-18/ 4:03 PM	02-03-18/ 2:27 PM	22 h and 24 min	Successful
Greens and browns (rice husk)	Green = 1 kilo Brown = 2 kilos	02-04-18/ 4:50 PM	02-05-18/ 5:12 PM	24 h and 22 min	Successful
Greens	3 kilos	02-05-18/ 8:24 PM	02-07-18/ 10:34 PM	50 h and 10 min	Successful
Browns set 1 (shredded)	3 kilos	02-10-18/ 2:04 PM	02-11-18/ 1:57 PM	23 h and 53 min	Successful
Greens and browns (shredded)	Greens = 1 kilo Brown = 2 kilos	02-11-18/ 8:47 PM	02-12-18/ 8:16 PM	23 h and 29 min	Successful
Browns set 2 (shredded)	3 kilos	02-21-18/ 8:21 AM	02-22-18/ 8:03 AM	23 h and 42 min	Successful
Browns set 2 (rice husk)	3 kilos	02-22-18/ 10:03 PM	02-23-18/ 9:37 PM	23 h and 34 min	Successful

Table 2 Conventional composting time for greens and browns

Component material	Quantity	Date/time started	Date/time completed	Number of h and min
Greens and browns (ice husk)	Green = 1 kilo Brown = 2 kilos	02-04-18/ 4:53 PM	02-21-18/ 10:02 PM	17 days, 5 h, and 9 min

alone has the fastest rate in completing the process and the greens alone have the slowest. Moreover, the 2:1 browns and greens ratio can also be considered fast because of the amount of material added that is only 2 h behind the browns alone. In comparing the shredded tree branches and rice husk, both have a close composting time. Thus, it can be considered that the shredded tree branches are as effective as the rice husk for brown materials.

The Table 2 shows that conventional composting for greens and browns takes 17 days, 5 h, and 9 min. In comparison with the current process, the proposed process is 94.10% faster.

$$\text{Proposed process} = 24\ \text{h and}\ 22\ \text{min} = 24.37\ \text{h} \tag{1}$$

$$\% = \left(|\text{current process} - \text{proposed process}| / \text{current process} \right) * 100 \tag{2}$$

$$= \left(|413.15\ \text{h} - 24.37\ \text{h}| / 413.15\ \text{h} \right) * 100 \tag{3}$$

$$= 94.10\% \tag{4}$$

4 Conclusion

The proponents were able to design and develop a bio composting machine that can accept shredded biodegradable materials and automatically mix them using a motor-driven mixer that promotes aeration during the process. The proponents were able to create a prototype that can monitor and maintain the temperature and monitor the moisture content of the compost material with the use of temperature sensor, blower as heater, and soil moisture sensor and were able to design and develop an automatic SMS notification regarding the completion of the process when the moisture content of the compost material reaches 18%. After the testing of the prototype, it was concluded that the reliability and accuracy of the prototype work continuously for more than 3 days in which it becomes successful without encountering any problems and the sensors used were calibrated with the help of a soil survey instrument which is used to compare the values that it measures to the values measured by the sensors, making it reliable and accurate. The prototype of the research was able to shorten the bio composting process using the proposed process which is much faster compared to conventional composting. In the proposed process, it only took 1–3 days to finish the composting process, unlike in the conventional and current process, which normally took more than a month to decompose the biodegradable wastes. Therefore, the proponents were able to prove in the study that the proposed system is an automated rapid bio composting with monitoring system.

5 Recommendation

For the future researchers, who have the courage and interest to study deeper and to know more about the process of bio composting with the help of automation and monitoring system, the proponents recommend the following for the betterment of the study.

It is recommended to include a shredding system into the bio composting machine to breakdown the biodegradable wastes. Another is to use an electric heater in exchange for the blower to increase the range of the temperature from 35 °C up to 60 °C without overheating the component and for a more stable heating operation. Also, consider adding a cooling system to automatically decrease the temperature of the compost material in case the temperature exceeds 60 °C.

To improve security, the proponents also recommend creating a user interface that allows the password of the prototype to be changed. It is also recommended to develop a data logger that can monitor and log data regarding the temperature and moisture content of the compost material which can be used for conducting experiments with new enzyme mixtures or such.

For the water supply, it is recommended to attach a level sensor or a float sensor to the water supply to check whether it is full or already running out of water. This will be displayed in the LCD for the user to be updated if it is needed to be refilled, as well as sending an SMS message to notify the user.

Further recommendations include research on some devices that allow checking of the remaining balance of the prepaid load for the SMS notification. A device that can indicate if there is currently a signal in the area can also help improve the study, as well as providing an internal power source or a battery for the prototype to sustain the operation in case of power interruption.

To improve the bio composting process, it is recommended to include the monitoring of carbon and nitrogen levels of the compost material along with its temperature and moisture content. Improving the size and the material used for the bio composting machine for stronger foundation and to improve its capacity is also recommended.

The recommendations are only based on the data gathered during the research and experiments made by the proponents of this study. Therefore, it is not guaranteed that these recommendations are attainable and can be used for the Bio Composting process until a proper experiment is conducted.

References

1. Cornell Waste Management Institute. (1996). Cornell University, Ithaca, New York USA.
2. Meenambal, T., Uma, R. N., & Saravannan, S. (2003). *Study on biodegradation of fruit waste aerobic composting*. Chennai.
3. Azuelo, M. C. C., Barbado, L. N., & Reyes, L. M. L. (2016). *Assessment of solid waste management strategies in Camarines Norte, Philippines*. Camarines Norte State College.
4. Onwosi, C. O. (2016). *Composting technology in waste stabilization: On the methods, challenges and future prospects*. Nsukka: University of Nigeria.
5. Palijon, A., Hara, Y., Murakami, A., De Guzman, C., & Yokohari, M. (2017). *Biowaste reuse through composting: The response of Barangay Holy Spirit in Quezon City, Philippines, to solid waste management*. Tokyo: Springer.
6. Wang, X. (2014). *Temperature and humidity monitoring system based on GSM module*. Dalian: Dalian Light Industry School.
7. Satish, R., & Pradeep, S. (2016). *Monitoring moisture of soil using low cost homemade soil moisture sensor and Arduino UNO*. Coimbatore.
8. Lohtia, S., James, W. M., & Chong, B. (2003). *System and method for providing subscriber-initiated information over the short message service (SMS) or a microbrowser*. Hwang.
9. Sundberg, C. (2005). *Improving compost process efficiency by controlling aeration, temperature and pH*. Uppsala: Swedish University of Agricultural Sciences.
10. E. Harshavardhan Goud, A. Harshika, G. Akhil, D. Charishma, K. Bhupathi, I. Kumara Swamy, Real-time based temperature control using Arduino, 2017.

Post-harvest and Processing Technology Management System for Local Coffee Growers

Willie C. Buclatin

1 Introduction

The Philippines is one of the few countries that produce four varieties of commercially viable coffee: Arabica, Liberica (Barako), Excelsa, and Robusta. Climatic and soil conditions in the Philippines—from the lowland to mountain regions—make the country suitable for all four varieties [1]. According to Tornincasa et al. (2010), coffee is the second most traded commodity worldwide after crude oil [2].

The Foundation for Sustainable Coffee Excellence founded by the premium coffee supplier Henry and Sons aims to address the major issues that are stopping coffee farmers from producing better coffee in larger quantities. The firm has been striving to improve the lives of coffee farmers, ensure local coffee industry's future, and ultimately, gain back Philippine coffee's prominence in the global market.

The Philippine Statistics Authority (2017) stated that the production of dried coffee berries dropped by 10.5%, from 22.59 thousand metric tons last year to 20.21 thousand metric tons. This decrease was caused by the following: (a) trees were toppled down and fewer berries were harvested due to strong winds brought by Super Typhoon "Lawin" in CAR (Kalinga) during flowering stage; and (b) prevalence of old and low-yielding trees in CALABARZON (Cavite) [3].

According to PEF (2016) on industry study on coffee, the country is 55% self-sufficient and 45% import-dependent, which is still far from the government's goal of reducing imports. In 2016, the total production of the four varieties of coffee was 137, 644 metric tons, while the total consumption reached to 180,000 metric tons. The local demand for coffee could increase to 100,000 MT in the coming years,

W. C. Buclatin (✉)
Department of Industrial Engineering and Technology, Cavite State University, Indang, Philippines
e-mail: willie.ceit@cvsu.edu.ph

A. Beltran Jr. et al. (eds.), *World Congress on Engineering and Technology; Innovation and its Sustainability 2018*, EAI/Springer Innovations in Communication and Computing, https://doi.org/10.1007/978-3-030-20904-9_4

but with local capacity at 25,000 MT annually, the country would never get out of coffee importation [4].

Besides genetic and environmental factors, the range of cares taken from field to cupping can affect coffee production, productivity, and finally coffee quality, costs, and revenue. Still there are gaps and lack of adequate information on the effects of post-harvest and processing technology techniques on coffee. This study was undertaken to identify and analyze the issues; challenges and problems encountered by local coffee farmers on post-harvest and processing technologies and integrate the appropriate post-harvest and processing technology management framework for local coffee farmers that need to be developed for an effective and sustainable coffee industry.

2 Methodology

This chapter presents the methodology used in determining the current coffee post-harvest and processing technology practices being observed by local farmers in the province of Batangas and Cavite. This includes the detailed description of the research design, target population and sample respondents, research instrument, data gathering procedures, validation, and administration, and the statistical treatment used in the analysis of the gathered data to find answers to the problems.

2.1 Research Design

This study was a descriptive type of research. Actual interview of the local coffee growers was conducted with the aid of a survey questionnaire in order to gather information and to generalize findings from a drawn sample.

This approach provides detailed profile of the local coffee farmers in terms of their demographic profile, farm characteristics, coffee farming practices, and technology management-related factors using the element of quantitative and qualitative approach. These are all important in designing a post-harvest and processing technology management framework in the province of Batangas and Cavite.

The approach employed in this study concerns with data gathering, recording, describing, analyzing, and interpreting the nature of the different variables in coffee post-harvest and processing technology management practices adopted by local coffee farmers in the province of Batangas and Cavite.

2.2 Population and Sample

The study was conducted in the province of Batangas and Cavite. The respondents in this study were local coffee farmers from nine (9) municipalities (Amadeo, Alfonso, Silang, Gen. Aguinaldo, Mendez, Tagaytay, Indang, Lipa, and San Jose) whose local coffee farmers are registered in the Office of the Provincial Agriculturist of both Cavite and Batangas.

The actual registered local coffee growers from the two recognized coffee producing provinces: Batangas and Cavite were obtained from the office of the Provincial Agriculturist. From a total population of 12,654 coffee farmers, 10,936 from Cavite and 1220 from Batangas, only 387 participants will be given a survey questionnaire.

Lists of local coffee growers were secured from the Office of the Provincial Agriculturist of both Batangas and Cavite.

2.3 *Sampling Technique*

Slovin's formula and stratified random sampling were used to determine the number of samples needed per municipality based on the total population of local coffee farmers in the province of Batangas and Cavite.

The number of participants from each province was determined using stratified sampling technique. With a total of 387 participants, 348 coffee growers from Cavite and 39 coffee growers from Batangas were determined.

Slovin's Formula

$$n = \frac{N}{1 + Ne^2} \tag{1}$$

where:

n = Sample size
N = Number of population
e = Margin of error (0.05)

Stratified Random Sampling Formula

$$n_\mathrm{h} = \left(\frac{N_\mathrm{h}}{N}\right) n \tag{2}$$

where:

n_h = sample size of the stratum h
N_h = population size of the stratum h
N = total population size
n = total sample size

After determining the number of samples per province, the stratified random sampling was also used to determine the number of samples per cities/municipalities based on the population of local coffee farmers. In order to conduct this sampling strategy, the researchers defined the population first. Coffee farmers from

Table 1 Distribution of participants per cities/municipalities of Cavite and Batangas

Cities/municipalities	Number of samples
Cavite	
Amadeo	150
Alfonso	55
Silang	42
General Aguinaldo	30
Mendez	30
Tagaytay City	21
Indang	20
Batangas	
Lipa	21
San Jose	18
Total	387

cities/municipalities were determined and listed to come up with the number of participants per area.

Table 1 shows the total number of participants from each city/municipality of Batangas and Cavite.

2.4 *Statistical Treatment of Data*

Gathered data were tallied, computed, and tabulated. Frequency counts and percentages were used to summarize and describe the overview of coffee industry in the Province of Batangas and Cavite.

In order for the researcher to determine the significant difference among the different techniques used by the farmers in various post-harvest processes in growing local coffee, the independent sample t-test was used. To determine if there is a significant difference, a level of confidence of 95% was used with $\alpha = 0.05$. This means that if the computed p-value is equal or below 0.05, the null hypothesis would be rejected. If the computed p-value is above 0.05, the null hypothesis would be accepted.

The formula for independent sample t-test is:

$$t = \frac{\overline{x}_1 - \overline{x}_2}{\sqrt{\frac{S^2}{n_1} + \frac{S^2}{n_2}}} \tag{3}$$

where:

$t = t$-value
$\overline{x}_1 =$ mean of first group

$\overline{x}_2$ = mean of second group
s^2 = pooled variance *(see formula below)*
n_1 = number of samples of first group
n_2 = number of samples of second group

The formula for the pooled variance is:

$$s^2 = \frac{\sum (x_1 - \overline{x}_1)^2 + \sum (x_2 - \overline{x}_2)^2}{n_1 + n_2 - 2} \tag{4}$$

where:

s^2 = pooled variance
x_1 = data point in first group
$\overline{x}_1$ = mean of first group
x_2 = data point in second group
$\overline{x}_2$ = mean of second group
n_1 = number of samples of first group
n_2 = number of samples of second group

A cost-benefit analysis was conducted and prepared. This helped to determine the appropriate coffee post-harvest practice and technology that will maximize the local farmer's production. Coffee sold in different coffee post-harvest technologies is the present worth of the benefits, while expenditures (i.e., maintenance of machine and wages of operators) are the present worth of the cost. If CBR >1, it indicates that it is recovering every amount of money invested. It was used to assess the viability of technology adapted by coffee farmers.

The productivity of the local coffee farmers was evaluated according to the rate of return on investment (ROI). It compared returns to costs by making a ratio from cash inflows and outflows that follow from the investment. By definition, the ROI ratio calculates as net investment gains divided by total investment costs.

A positive result means that returns exceed costs. Therefore, consider the investment a net gain.

The formulas are:

$$\text{Payment Period} = \frac{\text{Initial Investment}}{\text{Cash Inflow for the Period}} \tag{5}$$

$$\text{Profitablity Index} = \frac{\text{Present Value}}{\text{Initial Investment}} \tag{6}$$

$$\text{Net Present Value} = \text{Present Value} - \text{Initial Investment} \tag{7}$$

$$\text{Return on Investment (ROI)} = \frac{\text{Net Profit}}{\text{Total Investment}} \tag{8}$$

$$\text{Rate of Return on Investment} = \text{ROI} * 100 \quad (9)$$

$$\text{Profit} = \text{Sales} - \text{Cost} \quad (10)$$

The payback period is used in the study to evaluate the earnings per year of the local coffee farmers considering the collected data from the beginning of the project until the accrued incomes are equivalent to the cost of the asset, at which time, the outlay is said to have been paid back. In a certain investment, it will help local coffee farmers to answer their questions on how long does it take to get their investment back.

The net present value is described as the difference between the present value of the cash inflows and the present value of the cash outflows. Considering the results of the study, it is the project's present value less the initial investment of the coffee farmers. With this method, all cash flows are discounted to present values using the required rate of return.

The profitability index is an additional method which helps local coffee farmers in selecting the appropriate post-harvest and processing technology.

Considering the return of investment (ROI), it helps local coffee farmers analyze if their investment to post-harvest and processing technology could help achieve the goals.

For profit analysis, local coffee farmers could calculate if they are making a profit through the adoption of technology.

Lastly, for internal rate of return (IRR), local coffee farmers could evaluate the benefits or the attractiveness of adopting a post-harvest and processing technology.

3 Results and Discussions

3.1 Issues/Problems and Challenges Encountered by Local Coffee Farmers in the Province of Cavite and Batangas

In coffee post-harvest and processing, farmers experienced several problems that do affect the condition of the coffee production as well as their profit. These problems were identified as unfavorable weather condition, pests and diseases, high input price for post-harvest machines, lack of capital, insufficient machineries and technologies, and lack of manpower.

Table 2 shows the problems encountered by the local coffee farmers related to post-harvest technology. Insufficient machineries and technologies (72.21%) was the main problem experienced by Cavite and Batangas coffee farmers.

Table 2 Problems encountered by the Cavite and Batangas coffee farmers related to coffee post-harvest technology and processing

Problems encountered[a]	Frequency	Percentage
Insufficient machineries and technologies	291	72.21
Lack of capital	32	7.94
Pests and diseases	23	5.71
Lack of manpower	21	5.21
High input	20	4.96
Unfavorable weather condition	16	3.97

[a]Respondents provide multiple answers to the question

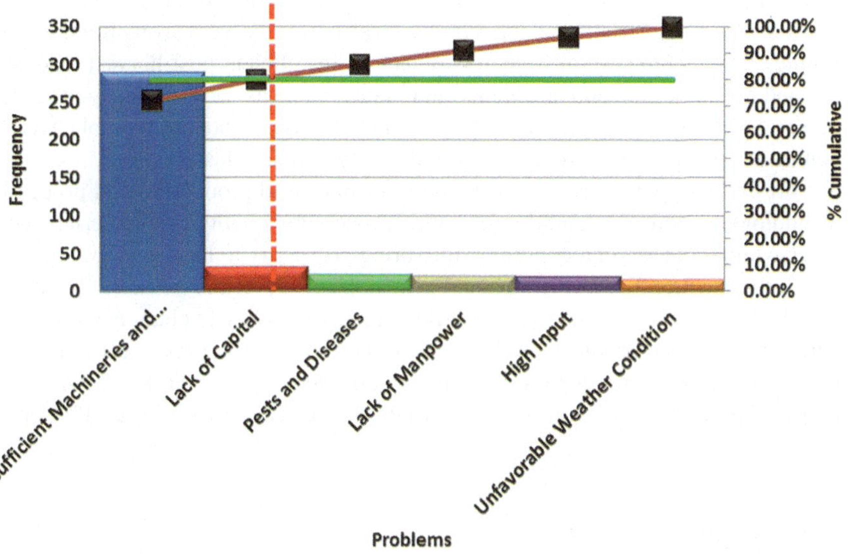

Fig. 1 Pareto analysis of the study

3.2 *Pareto Analysis*

The Pareto analysis and the results of the study focused on the cause which generates the problems. This chart helped find the majority of the problems and their root causes. The results of the Pareto were based on the problems encountered by the local coffee growers and are presented in Table 14. The Fig. 1 above shows that insufficient machineries and technologies 72.21% was the major problem being encountered by the local coffee farmers in the province of Batangas and Cavite. This is in agreement with the results of the survey conducted that a majority of the local coffee farmers are using manual technology for coffee processing. Since, they are only small processors, the local coffee farmers pointed that they lack government support, they lack proper maintenance of the existing machineries and they lack the needed knowledge and technical know-how.

3.3 SWOT Analysis of the Study

The internal (strengths and weaknesses) and external (opportunities and threats) factors that have an impact on the viability of coffee post-harvest and processing technology were determined.

The information presented provides a great help in the capabilities of local coffee growers to the competitive environment and it is instrumental in strategy formulation and selection.

Relative to its strength, the coffee varieties could be grown throughout the country. The Philippine coffee is known and traded worldwide; there is a growing demand for coffee, and it is considered as one of the basic consumer needs.

Contrary to the results of the survey, local coffee farmers from Batangas and Cavite encountered several weaknesses such as insufficient post-harvest machineries and technologies and problems and issues such as lack of manpower, lack of capital, high input, presence of pest and diseases, and unfavorable weather conditions that possibly hinder the growth of the coffee industry.

Part of the opportunities are availability of improved production and processing technologies, increasing number of coffee processors, quality raw materials, and a diverse target market that could possibly bring prosperity to local coffee farmers.

Lastly, local coffee farmers should always be ready of the possible threats that could affect coffee production. The following threats (climate change, strong competition, land conversion, increasing health consciousness, presence of substitute products, and aging producer population and limited engagement by youth) should be given proper attention in order to sustain the coffee industry in the Philippines (Table 3).

Table 3 SWOT analysis of the study

Strengths	*Weaknesses*
All four varieties of coffee could be grown throughout the country Philippine coffee is known and traded worldwide Growing demand for coffee shops Coffee is a basic consumer need Most of the farmers were the owner of the coffee farm	Inadequate post-harvest machineries and technologies Low level of educational attainment of the participants Problem/issues encountered by local coffee farmers such as lack of manpower, lack of capital, high input, pest and diseases, and unfavorable weather conditions
Opportunities	*Threats*
Availability of improved production and processing technologies Increasing number of coffee processors, quality raw materials Diverse target market	Erratic agro-climatic condition/ climate change Stiff competition Land conversion Increasing health consciousness Substitute products Aging producer population and limited engagement by youth

3.4 Inventory of Current Post-harvest and Processing Technology Adopted by Local Coffee Farmers in Batangas and Cavite

Table 4 shows the matrix of coffee post-harvest and processing technology adopted by local coffee farmers in Batangas and Cavite.

3.5 Element of a Proposed Coffee Post-harvest and Processing Technology Management Framework

The Philippine coffee industry is under the collective force to improve the quality of appropriate coffee post-harvest and processing technologies that will give benefits to coffee players. The adoption of appropriate coffee post-harvest and processing technologies has become very timely to answer the needs of the local coffee farmers in the Philippines.

3.6 Economic Feasibility

In general, the agricultural condition in the Philippines is a priority area considering that there are already available technologies that will bridge what is already available in more advanced progressive areas as compared to those who are located in provincial areas where connectivity is a major problem.

Data being gathered was analyzed considering different techniques on coffee post-harvest and processing technology used by local coffee farmers through cost-benefit analysis.

3.7 Cost–Benefit Analysis (Profit and Loss Analysis)

Farmers used different practices and technologies in coffee processing. Through the data gathered, it shows that several farmers invested in some machinery such as hulling machine, sorting machine, roasting machine, and grinding machine. With

Table 4 Coffee post-harvest and processing technology adopted in the Philippines

Process	Machineries	Current technologies adopted
Picking	Manual	Selective picking, stripping
Drying	Coffee dryer	Sun-dry, drying machine
Hulling	Coffee huller	Manual hulling, hulling machines
Sorting	Coffee sorter, 12″ square wood sizing screen	Manual sorting, sorting machines
Roasting	Coffee laboratory roaster	Manual roasting, roasting machine
Grinding	Coffee grinder	Manual grinding, grinding machines

the use of ROI, costs of return are identified in percentage whether the opportunities are worth to be invested.

The result shows that selective picking is much viable in the picking process based on the cost per production of 1:19 compared to stripping of 1:28 ratio considering the gathered data. This also means that for every production, only 19 pesos production cost shall be incurred under selective process which gives a higher quantity of production with less number of days and workers.

The analysis also focused on stripping to sun drying process where local coffee farmers sell the coffee berries to the market. Stripping to sun drying process is more viable based on the percentage of profit to sales of 57% as compared to 54% of selective to sun drying. The analysis is based on the percentage of performance of cost and profit to sales. This examines the decision in terms of its consequences or costs and benefits.

Results revealed that as to profits, manual hulling generates 80% based on sales, while hulling machine generates 67% based on sales.

As to cost, manual hulling incurred 20% cost based on sales and hulling machine incurred 33% cost based on sales in which 44% of its total cost came from depreciation of the machine which is 14% out of 33%.

Comparing the results of profit and loss analysis, manual hulling generates high percentage of profits as compared to machine hulling, but machine hulling is still viable in the process because of the 14% depreciation from the total percentage of cost that shall form part of fixed asset of the owner.

Payback period was also considered when it comes to the adoption of the hulling machine. Considering the results of the analysis, the payback period is 4.42 years. The result shows that in year 5, the cumulative cash flow sign changes from negative to positive, meaning that at a certain point between year 4 and 5, costs would be recovered by generated profit.

Profitability index analysis was also considered in the study. Based on the results of the analysis, the profitability index is 1.48; this means that the project creates value. It generates a return greater than the required return and the project has a big impact on the project's net present value.

As to profits and cost analysis, using sorting machines in the process is viable compared to the manual process.

Considering the results of the analysis, the payback period is 0.65 years. The costs would be recovered (first year) by generated profit.

Similarly, based on the results of the analysis, the profitability index is 4.26; this means that the project creates value. It generates a return greater than the required return and the project has a big impact on the projects net present value.

The result of the study also shows that as to profits and cost analysis in the roasting process, using roasting machines in the process is viable compared to the manual process.

And the payback period is 3.22 years, meaning that at some point between year 3 and 4, costs would be recovered by generated profit. So, the payback period is somewhere in the third year and the profitability index is 1.7; this means that the project creates value. It generates a return greater than the required return, and the project has a big impact on the project's net present value.

As to profits and cost analysis for grinding process, the use of grinding machines in the process is viable compared to the manual process with a payback period of 1.76 years and a profitability index of 2.2, which means that the project creates value.

3.8 Post-harvest Processes Matrix in Terms of Cost–Benefit Analysis (Profit and Loss Analysis)

Table 5 shows the matrix of the post-harvest and processing technology for local coffee growers. The matrix was based on the results of the cost-benefit analysis done on the different techniques used by the farmers. The results were evaluated based on the profit and loss analysis.

3.9 Coffee Post-harvest and Processing Technology Management Framework

Quality function deployment Presented in Fig. 2 is the quality function deployment for the coffee post-harvest and processing technology. The customer requirements involve provision of machineries, innovation of existing manual, strong access to financial and funding institutions, providing funding assistance to the local coffee farmers from the government, and reduction of inputs. Meanwhile, prioritization matrix was identified considering the technical requirements and the set of priorities (business networking, benchmarking with other coffee producing localities, adoption of technology, strengthening the linkage of coffee SMEs through financing from other funding agencies, coordination of coffee association to the academes through research and development, industry updates, and attending investment and financing forum on coffee plantation). The relationship of the requirements was identified as strong relationship (9), medium relationship, and (3) weak relationship (1). The hose of quality ensures the accurate deployment

Table 5 Matrix for post-harvest technology practices for local coffee growers in the Philippines

Post-harvest processes	Best practices
Picking	Selective picking
Drying	Sun drying
Hulling	Hulling machine
Sorting	Sorting machine
Roasting	Roasting machine
Grinding	Grinding machine

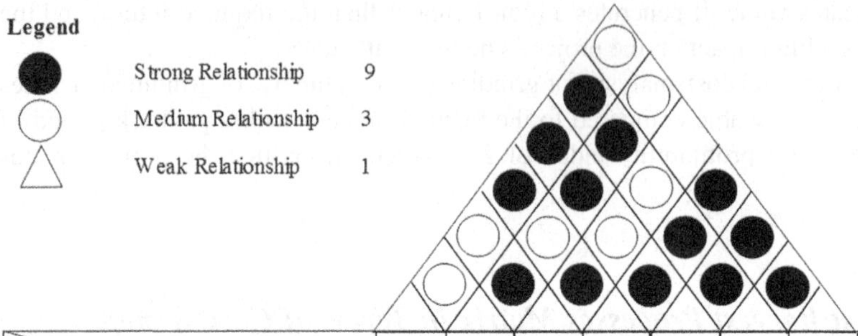

Technical Requirements / Customer Requirements	A	B	C	D	E	F	G	Customer Rating
Provision of machineries	△		●	●	○		○	5.00
Innovation of existing manual machineries	○	●	●	●	○	○	○	5.00
Strong access to financial and funding instutions	●		●	●	●	△	●	4.00
Providing funding assistance to the local coffee farmers from the government	●		●	●	●	△	●	4.00
Reduction of inputs	△	●	●	○	△	△	△	3.00
TARGET VALUES	**95**	**72**	**189**	**171**	**105**	**26**	**105**	
FACTORS CHOSEN FOR DEPLOYMENT			√	√	√		√	

Fig. 2 House of quality

throughout the process from planning up to monitoring of the post-harvest technologies. The result shows that the technical requirements to be deployed were adoption of technology, strengthening linkages of coffee SMEs through financing from other funding agencies, coordination of coffee association to the academes through research and development, and attending investment and financing forum on coffee plantation leading to a sustainable and profitable coffee industry.

Legend for technical requirements:

A – Business networking
B – Benchmarking with other coffee producing localities
C – Adoption of technology
D – Strengthening linkages of coffee SMEs through financing from other funding agencies
E – Coordination of coffee association to the academes through research and development
F – Industry updates
G – Attending investment and financing forum on coffee plantation

3.10 Significant Relationship of Post-harvest Processing Technology to Cost, Income, and Profit

Post-harvest technology matrix practices Table 6 shows the matrix of the post-harvest and processing technology for local coffee growers. The matrix was based on the results of the statistical tests done on the different techniques used by the farmers.

The results were evaluated based on the significant differences of the techniques in terms of the cost incurred, income generated, and profit realized by the farmers.

Based on the statistical analysis done in the picking process, there was no significant difference between the picking techniques used by the farmers. In terms of cost, both selective picking and stripping could be used since there are no sufficient evidences to support that there is a significant difference between the two techniques.

Generally, the farmers are using the sun drying technique in drying their coffee; hence, only the sun drying technique was included in the post-harvest processes matrix. Only the income and profit are considered in determining the appropriate practice for the process since the cost from the picking stage is carried over to drying.

In terms of cost, manual hulling was the technique recommended since there was a significant difference between the two practices used by the farmer in hulling. Manual hulling incurred less cost as compared with the use of a hulling machine.

In terms of income and profit, the use of hulling machine was the technique recommended since it generates higher income and profit.

Based on the statistical test done, it was found that these differences in income and profit are both significant statistically. Overall, since both income and profit greatly outweighed the cost, it is recommended that farmers should use hulling machines.

Table 6 Summary for post-harvest technology practices and operations for local coffee growers

Post-harvest and processing technology	Cost	Income	Profit	Overall
Picking	Selective picking and stripping	N/A	N/A	Selective picking and stripping
Drying	N/A	Sun drying	Sun drying	Sun drying
Hulling	Manual hulling	Hulling machine	Hulling machine	Hulling machine
Sorting	Manual sorting	Sorting machine	Sorting machine	Sorting machine
Roasting	Manual roasting	Roasting machine	Roasting machine	Roasting machine
Grinding	Manual grinding	Grinding machine	Grinding machine	Grinding machine

Manual sorting incurred less cost to farmers based on the mean cost when the two techniques were compared. Based on the statistical test, it was found that there was a significant difference between the two practices in terms of cost.

On the other hand, in terms of income and profit, the use of sorting machine is recommended since it was established that there is a significant difference between the two techniques used by the farmers.

It is recommended that farmers use sorting machine in their post-harvest processes since the benefits of the income and profit is more than the cost saved.

For the roasting process, the recommended technique in terms of cost is the use of manual roasting. Statistically, there is a significant difference between the two techniques used by the farmers.

However, the use of roasting machine is recommended in terms of the income and profit realized. In terms of cost, it is recommended that the farmers use manual grinding. The cost incurred is less as compared to the use of a grinding machine.

It was also proven that the differences between the two techniques are statistically significant.

On the other hand, in terms of income generated and profit realized, it is recommended that farmers use a grinding machine. Moreover, the difference of technique is statistically significant as compared to the use of manual grinding.

3.11 Proposed Coffee Post-harvest and Processing Technology Management Framework

Coffee post-harvest and processing development framework Figure 3 presents the coffee post-harvest technology development framework that puts together all the elements necessary in attaining a profitable and sustainable coffee industry leading to improved socioeconomic well-being of the local coffee farmers particularly in Batangas and Cavite.

The intervention considering the use of technology shall address priority concerns along with the support services and training that facilitates the implementation and adaptation and/or utilization of coffee post-harvest and processing technology.

Marketing support and policy environments are also important factors that ensure successful implementation of identified technologies programs and other interventions.

DA, LGUs, farmers' organization, and other NGOs, funding institutions, Philippine Coffee Board, DOST, DOLE, DENR, academe, and International Coffee Organization must collaborate in identifying the coffee post-harvest and processing technology priority concerns that are need-specific, and task-specific. Also, the appropriate coffee post-harvest and processing technology and appropriate interventions shall be carefully identified by the stakeholders so that financial and

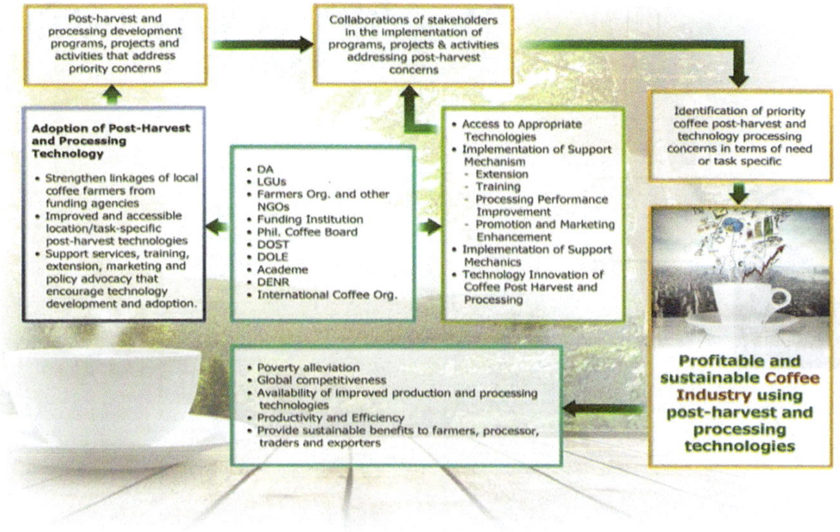

Fig. 3 Coffee post-harvest development framework

human capital investments are used effectively and efficiently. Thus all efforts lead to a profitable and sustainable coffee industry and improved well-being of coffee farmers. Hence, the ultimate goals of poverty alleviation; availability of improved production and processing technologies; productivity and efficiency, providing sustainable benefits to coffee farmers, processors, traders, and exporters; and global competitiveness could be realized.

3.12 *Coffee Post-harvest and Processing Technology Adoption and Transfer Management Framework*

As shown in Fig. 4, the strategies in the process decision plan were adopted from the management function such as planning, organizing, staffing, directing, and controlling.

The planning phase consists of the brainstorming on strategic direction and action programs on major issues, setting of coffee industry mission and vision, and drafting of management plan. When the plans are accomplished, the data and information will be organized. Here, strategies are made in order to achieve the initial goal. Activities will be performed like periodic meeting of the coffee technical working group, conducting regional industry profiling on coffee, strategic planning at all levels, aligning and participation to the regional development plans project for coffee, benchmarking analysis of the coffee industry, providing information

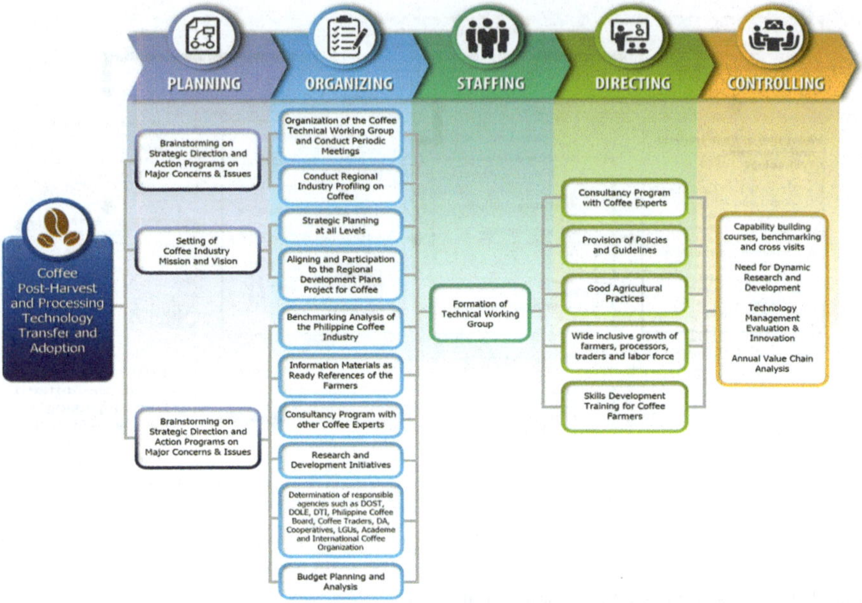

Fig. 4 Coffee post-harvest and processing technology adoption and transfer management framework

materials as ready references to the farmers, consultancy programs with other coffee experts, research and development initiatives, determination of responsible government and private agencies, and conducting budget planning and analysis. To be able to put the strategies into action, a technical working group will be formed in the staffing phase and will undergo several trainings, activities, and programs for them to grow and perform their tasks properly. Lastly, there should be a continuous monitoring and controlling in order to improve the standards of coffee processing leading to a sustainable and profitable coffee industry.

3.13 *Five M's Sustainability Management Plan in Accordance to the Proposed Framework*

While sustainability is about the future of the society, for today's industries and businesses, it is also about commercial success.

The matrix presented below identifies the sustainability factors and sustainability objectives that projects need to target in order to build the potentials for a sustainable development (Table 7).

Table 7 Sustainability management plan

Sustainability factors	Sustainability objective	Sustainability actions
Manpower	Educate local coffee farmers with regard to post-harvest and processing technology	Skills development trainings Skills training and seminars on coffee processing Conduct demonstrations and insist the possible benefits of adopting the new technology Conduct regional local coffee farmers profiling Conduct periodic meetings
Method	Provide strategies for continuous improvement on post-harvest and processing technology, and improve the quality of coffee-appropriate post-harvest technologies	Organization of the technical working group Benchmarking on coffee processing and cross visits Provide information materials that will be distributed as ready references to the farmers Partnership with LGUs, DA, DENR, farmers' cooperative, academe, Philippine Coffee Board, DOST, and other government agencies Trade fairs and exhibits Market-matching activities Improve packaging and label design Consultancy program with international coffee experts Conduct dynamic research and development Conduct strategic planning at all levels Standardization of coffee post-harvest technology and processing Perform annual value chain analysis Conduct periodic meetings to evaluate the technology management
Machine	Analyze and adopt the technology requirements	Create a technical working group that will focus on the analysis of technology requirements Benchmarking and cross visits Perform research and development Design and development of appropriate machineries for post-harvest technologies Evaluate the post-harvest technologies

(continued)

Table 7 (continued)

Sustainability factors	Sustainability objective	Sustainability actions
Money	Provide sufficient funds for local coffee farmers to sustain the post-harvest technologies	Strengthen linkage of coffee farmers through financing for and directory of financial packages Determination of responsible agencies such as DOST, DOLE, DTI, Philippine Coffee Board, Coffee Traders, DA, Cooperatives, LGUs, academe, and International Coffee Organization Budget planning and analysis
Management and control	Develop and conduct a comprehensive project evaluation	Design and implement a comprehensive evaluation plan to capture the effectiveness of the project Conduct a technology management evaluation through research and development Continuous seeking for possible funding agencies for a technology management sustainable development

3.14 Coffee Post-harvest and Processing Technology Roadmap

A coffee post-harvest and processing technology roadmap was developed which will serve as a planning technique to support the coffee post-harvest and processing technology adoption and transfer management framework with specific technology transfer solutions. The said roadmap may help the government and other policy makers leading toward a sustainable coffee industry (Fig. 5).

The strategies focused on technology development and technology adoptions will be of great help to increase the local coffee farmer's productivity and production of coffee leading to self-sufficiency.

Considering a positive change particularly on the production of coffee, it will help improve the farmer's standard of living through diversified high value agriculture.

There will be a great impact particularly on the socioeconomic aspects, for it will increase rural employment.

The emergence of technologies particularly on coffee post-harvest processing will surely lessen coffee bean and coffee products importation; improve income of farmers, processors, and other stakeholders; result in a more profitable coffee industry in the value chain; and would continuously improve the quality of coffee post-harvest and processing technologies for a sustainable coffee industry that is self-sufficient with a significant share of the export market.

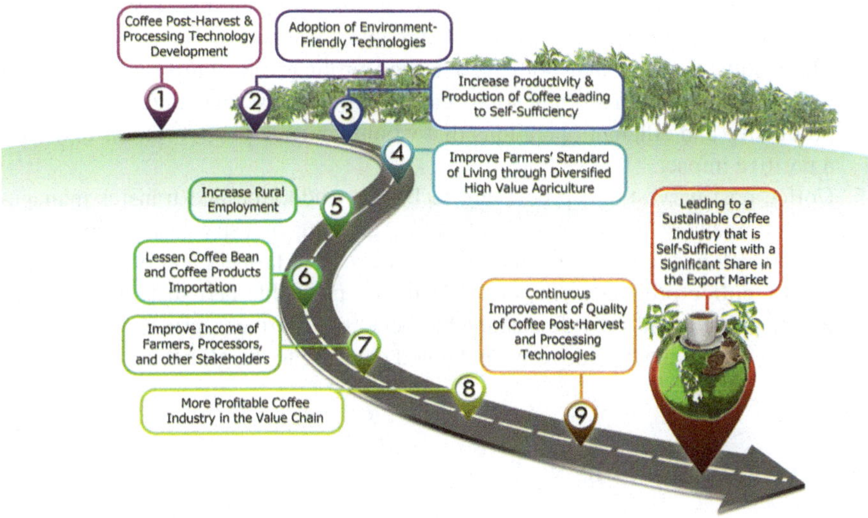

Fig. 5 Coffee post-harvest and processing technology roadmap

4 Conclusion

Based on the results of the study, the following conclusions are drawn:

1. The demographic profile and farm characteristics of the farmers showed that not only men are engaged in farming activities but also includes women. Mostly, the ages of these farmers range between sixty (60) and above. Most of the respondents are high school graduates. In addition, majority of the farmers owned the land they are farming. Thus, they have the experience in cultivating, harvesting, and processing coffee.
2. In general, most of the farmers have experienced several issues, challenges, and problems pertaining to coffee post-harvest and processing. Among the issues and challenges faced by local coffee farmers are unfavorable weather conditions, high input, lack of manpower, pests and diseases, lack of capital, and insufficient machineries and technologies which rated to be the highest with a percentage of 72.21%. Farmers from different municipalities of Batangas and Cavite are not capable of procuring their own coffee post-harvest and processing technology; thus, majority of the local coffee farmers are still practicing the traditional or conventional operations. This can be attributed to the low or limited access to financial or funding institutions, hence, resulting in limited machineries and technologies.

3. Farmers generally adopted the manual process in picking (selective and stripping) coffee berries. The use of machineries for a productive and effective operations or adopting the technologies creates value and has a big impact on the farmers considering the process that requires technology.
4. The results revealed that both income and profit greatly outweighed the cost of investment; it is recommended that the farmers use machineries to ensure productive and efficient operations that would generate greater returns and have a positive impact.
5. Coffee post-harvest and processing technology adoption and transfer management framework on the growth of coffee growing industry in the Philippines were developed to maximize the profit of the coffee farmers and provide higher quality coffee products. This could lead to a profitable and sustainable coffee industry that could provide sustainable benefits to farmers, processors, traders, and exporters, alleviate poverty, and could compete to the global arena.

5 Recommendations

The following recommendations are hereby suggested based on the findings and conclusions of the study.

5.1 Technology Adoption

To use the proposed coffee post-harvest and processing technology and transfer management framework, there is need for the government to provide training and development and instill the importance of adopting coffee post-harvest and processing technology to enhance retention of coffee leading to a sustainable coffee industry.

5.2 Education, Seminars, and Skills Development Training for Coffee Farmers

To augment the productivity and sustainability of coffee industry, farmers need to be trained on entrepreneurial skills so that they can invest their farm profits into more income-generating assets like post-harvest machineries so as to harness more farming capital. They should be provided with more extension services, seminars, and trainings about correct handling and processing regarding appropriate technologies.

5.3 Further Studies

To further enrich the findings and the adoption/transfer of technology, there is a need for further research and development or technology management evaluation as part of monitoring and control for a sustainable coffee industry. Supplementary or similar study be conducted in other coffee-producing areas in the country to determine the optimal set-up of the post-harvest and processing technology practices for coffee in the country as a whole. Also, consider other crops if the framework is also applicable.

Acknowledgments First and foremost, the author would like to give his highest honor and praise to our Almighty Father for bestowing us the gift of life, sanity, and enough courage and strength to continue and accomplish this dissertation. The author would like to acknowledge all those who generously shared their knowledge and time so generously and those who have selflessly helped and contributed in so many ways.

References

1. Philippine Coffee Board. (2013). *Philippine coffee*. Retrieved September 2, 2016, from http://philcoffeeboard.com/philippine-coffee/
2. Tornincasa, P., Furlan, M., Pallavicini, A., & Graziosi, G. (2010). Coffee species and varietal identification. In *Tools for identifying biodiversity: Progress and problems* (pp. 307–313). EUT. ISBN: 978-88-8303-295-0
3. Bersales, L., Perez, J., & Recide, R. (2017). *Philippine Statistics Authority*. Quezon, Philippines: Major non-food ands industrial crop, *11*(1), ISSN: 2094-6198.
4. PEF. (2016). *A primer on PEF's priority commodities: Industry study on coffee* (pp. 13–14). Retrieved from http://8liens.com/pef.ph/wp-content/uploads/2016/03/Industry-Study_Coffee.pdf

Technology of Obtaining Long-Length Powder Permeable Materials with Uniform Density Distributions

Oleg Zabolotnyi, Oleksandr Povstyanoy, Dmytro Somov, Viktor Sychuk, and Kostiantyn Svirzhevskyi

1 Introduction

One of the basic tasks in the production of materials by methods of powder metallurgy is to obtain products with a uniform density distribution. The heterogeneity of materials depends on many factors: the method of compaction, the shape and size of the details, the size and shape of the powder particles, ability to compact, the quality and quantity of lubricants, the speed of pressing, etc. Along with the gradient of density, the heterogeneity of mechanical strength, hardness, electrical conductivity, magnetic and gas permeability, and, also, distortion of the shape of the detail during the pressing and sintering is considered [1]. The issue of the distribution of stresses and, as a result, the density in the volume of briquettes is still not sufficiently studied. The development of new measuring devices provides an opportunity to experimentally estimate the density distribution and porosity in the volume of the finished product.

In modern technology, interest in highly porous materials and materials with regulated porosity is increasing [1–3]. The basic parameters of these materials, which determine their suitability for use in workpieces, are porosity and high permeability in relation to liquid and gas media. The porosity and uniformity of their volume distribution are the basic indicators of the possibility of their efficient use for the implementation of processes such as filtration, dispersion of gases and liquids, catalysis, fire retardation when igniting gas mixtures, sound deadening, etc. [2, 3].

O. Zabolotnyi (✉) · O. Povstyanoy · D. Somov · V. Sychuk · K. Svirzhevskyi
Lutsk National Technical University, Lutsk, Ukraine
e-mail: volynasi@gmail.com; povstjanoj@ukr.net; somovd@rambler.ru; svamator@gmail.com; kostonmen@gmail.com

A. Beltran Jr. et al. (eds.), *World Congress on Engineering and Technology; Innovation and its Sustainability 2018*, EAI/Springer Innovations in Communication and Computing, https://doi.org/10.1007/978-3-030-20904-9_5

With powder permeable materials (PPM), the most widely used technique is the obtaining of filtering materials (FM). Cleaning techniques based on filtration through porous elements are increasingly being used in new technologies to ensure the reliability of machines by improving the cleanliness of lubricants and fuels, improving purity, and improving product qualities. The quality and lifetime of such PPM are determined by the uniformity of their porous structure [1].

2 Literature Review

Among the various metal and material processing methods, powder metallurgy has a special significance, because it not only allows to obtain products of various shapes and purposes, but also allows to create basically new materials with special and often unique features, structure, and composition. It is difficult and often impossible to obtain such materials by other methods. More than any other operation, the process of pressing (forming) limits the technological capabilities of powder metallurgy and determines not only the performance characteristics of the finished product, but also reveals hereditary influence on a number of the most important physical and mechanical properties. The non-uniformity of the density distribution in powdered briquettes after pressing is often manifested at the stage of sintering in the form of warping the sample, the formation of micro- and macrocracks, or its uniform destruction. This results in nonuniform mechanical, filtering, electrical, and other properties of the product [1].

Of all the variety of powder products, long-length products should be distinguished, the need for which is constantly increasing in connection with the development of new technologies. Currently, powder metallurgy faces some difficulties in the production of this class of products. Traditional methods of making such materials have several advantages and disadvantages. The choice of method for forming parts from powders is determined by the properties requirements of the product, raw materials, energy costs, type of production, etc. [1–3].

An optimal scheme for obtaining a wide range of powder materials with a significant ratio of length to transverse dimensions (tubes, rods, products of complex forms) is the radial scheme of pressing [1, 4]. It is expedient to implement this scheme with the help of isostatics, which solves the problem of the production of durable products, provides the necessary quality, has insignificant labor costs, permits production safety, and enables the possibility of being utilized in different types of production [1, 4]. In this case, it is possible to control the quality of products and mechanize and automate the processes of pressing by applying relatively simple equipment, fittings, and tools.

However, along with the undeniable advantages of using dry radial-isostatic pressing (DRIP) to obtain long-length PPM, this technology has a number of significant constraints: the problem of obtaining powder materials with uniform density distribution by volume, high labor costs involved in instrument making and low reliability and service life, low automation of the process, and the inflexibility of production [4, 5].

Therefore, further study of the features of the DRIP process and study of powder materials of various shapes and sizes obtained by this method will allow us to determine the causes of the formation of zones of inhomogeneous density distribution (porosity) according to the volume of the product. This will enable us to develop a set of measures to eliminate the existing limitations in the manufacture of long-length powdered products by improving existing equipment, fittings, and tools, and developing a more efficient technology for obtaining high-quality PPM of simple and complex forms with uniform structural and operational characteristics.

3 Research Methodology

3.1 Improvement of Equipment for DRIP of Long-Length PPM

The elastic tools (insertions, shells, corks), which are located in the press block, play a main role in the pressing process. A set of requirements aimed at ensuring the reliability of work, technological design, and providing the necessary properties of PPM has been proposed. For the DRIP process and the study of its energy and power parameters, we have designed and manufactured an advanced experimental installation that allows work in static load modes [5]. The design of the press block of the DRIP machine and its general appearance are presented in Fig. 1. In the basic version, the installation for the DRIP process is as described previously [1], which allows receiving products from compaction materials of different nature and shape with maximum dimensions of $L \leq 320$ mm, $D \leq 60$ mm, and $p \leq 150$ MPa.

The Principle of Press Block Work Between the hull (1) and the elastic insertion (2) is an area with a working fluid. Before inserting the form into the press block, the gap between the elastic shells (6) and mandrel (7) on the vibration stand is filled with powder (11), after which the form is closed by the shim (8) and fixed by the nut (9). The assembled form is installed in the inner cavity of the insertion (2). After the shutter (12) is screwed in, in order to limit the axial deformation, we can start the radial compression of the powder pipe. This is carried out by creating pressure in the installation of high pressure and transmitting it into the working cavity of the press block through the fitting (13). The control of the working pressure is carried out by a manometer (14), which is installed in the hull (1). After finishing the pressing process, the form with powder briquette is removed from the press block and the process is repeated.

In the basic version of the press block [1], the authors used a polyurethane insertion with cuffs that were made by cutting. When machining on the surface of the insertion and the high-pressure cuff, micro-inequalities are formed (micro-damages), the quantity and magnitude of which depend on the physical and mechanical characteristics of the machining materials and cutting regimes. These

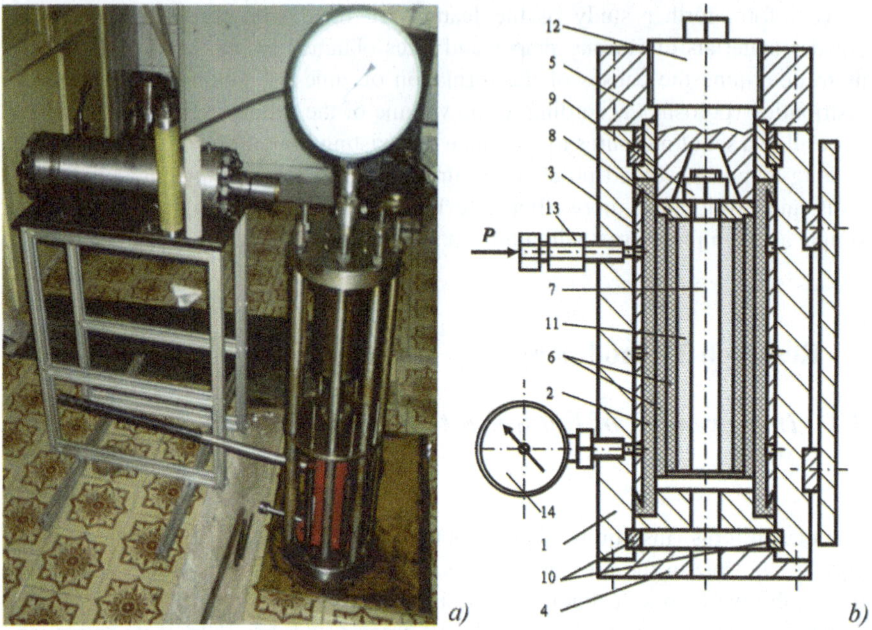

Fig. 1 (**a**) General view of the traditional (basic) installation for the dry radial-isostatic pressing (DRIP) method [1] and (**b**) improved design of the press block [5]: 1 – hull; 2 – elastic insertion (polyurethane); 3 – reinforcing element (steel); 4, 5 – covers; 6 – elastic shells (set); 7 – mandrel; 8 – shim; 9 – nut; 10 – cutter cams; 11 – powder; 12 – shutter; 13 – fitting; 14 – pressure gauge

micro-inequalities are stress concentrators and greatly reduce the durability of the elastic insertion, especially when cyclic loading is what we have in the DRIP process. As a result, the stability of the press block insertion was up to 1000 load cycles [1, 4]. The cost of polyurethane is high, so there is a need to improve the design of the elastic insertion of the press block to increase its reliability, machinability, and reduce the cost of the main elements of the press block. The improved structure of the press block (see Fig. 1) differs from the existing analogy by the fact that, inside the body (1), a special reinforced elastic insertion (2) with a steel reinforcing element (3), made according to the technology developed by us, is installed [5].

Figure 2a shows the traditional design of an elastic insertion with high-pressure cuffs that were machined with cutting [1]. The above construction has significant drawbacks. The cuffs of the elastic insertion are made with tooling with the help of a specially shaped cutter, which leads to the emergence of macro- and micro-damages in the working area of the cuff and significantly reduces the reliability of its work. In addition, the need to manufacture a cuff by cutting leads to additional costs for the manufacture of a special cutting tool (shaped cutter) and increases the volume of the intermediate medium, which causes additional energy costs for compressing the working fluid [1].

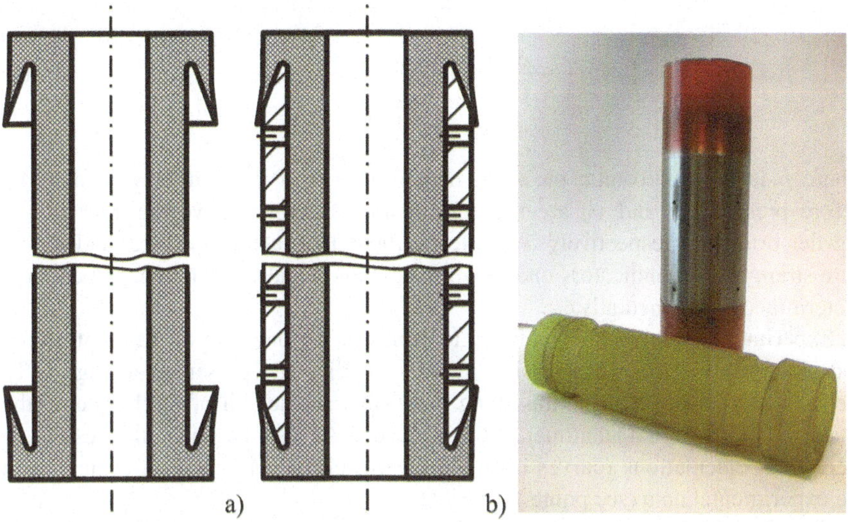

Fig. 2 (**a**) Construction of a traditional polyurethane elastic insertion obtained by cutting; (**b**) improved design of the reinforced elastic insertion; (**c**) general view of traditional and reinforced polyurethane elastic insertions

Figure 2b represents the improved design of the reinforced elastic insertion [5] characterized by the presence of a reinforced steel insertion, which reduces the volume of the intermediate medium and, simultaneously (at the stage of its casting), forms the high-pressure cuffs. Use of the casting method in the manufacture of the insertion allows avoiding the negative effects of the cutting process. Its constitutive feature is the presence of a steel element within it, which allows to improve the technological design, reduces labor required for the formation of high-pressure cuffs, increases the reliability and durability of the press block, and, also, reduces the volume of intermediate media that transmits pressure. The general views of a traditional elastic insertion with cuffs obtained by cutting [1] and an improved design of the reinforced elastic insertion [5] are presented in Fig. 2c.

3.2 Investigation of the Properties of the Produced Long-Length PPM and Improvement of the Pressing Schemes

The DRIP process of powder pipes was performed according to the well-known classical technology [1] by compaction of the powder to the mandrel in the isostat with the restriction of longitudinal deformation (see Fig. 1). To calculate the required pressing pressure by the radial scheme and to ensure the required porosity of the PPM, Eq. (1) was used [5]:

$$p = 2\sigma_{T_0}\left(\frac{\upsilon}{\upsilon_0}\right)^m \ln\frac{1-\upsilon_0}{1-\upsilon}\sqrt{1+\frac{\left(\frac{\upsilon}{\upsilon_0}\right)^{2l}(1-\xi)^2}{9\left(\frac{\upsilon}{\upsilon_0}\right)^{2m}-(1+2\xi)^2\left(\frac{\upsilon}{\upsilon_0}\right)^{2l}}} \quad (1)$$

where p is the hydrostatic pressure; σ_{T0} is the yield limit of powder material before pressing; υ and υ_0 are the current and starting relative densities of the powder briquette, respectively; ξ is the coefficient of lateral pressure; and m and l are strengthening indicators under hydrostatic and axial compaction, respectively (determined experimentally).

Experimental curves of the ability to compact for powders of different nature and size during radial compression in static loading are presented in Fig. 3. To measure the density distribution along the radius and length of the briquette, the imaging microscope Quantimet 720 was used. As can be seen, the results of theoretical calculations (curves obtained by Eq. (1)) are in good agreement with the experimental data (see points).

Figure 4 shows the experimental data of the density distribution dependence on the radius of the tubular powder briquette from the titanium powder of the mark PTS-1 with finite dimensions ∅44 × ∅32 mm. As can be seen from the dependencies, there is a slight change in density within 1–1.5% along the radius of the tubular powder briquette. This gradient of density, due to its insignificant magnitude, does not significantly affect the operational properties of the obtained PPM. Therefore, in the future, we believe that, for the thin-walled PPM, the density over the radius of the briquette is uniformly distributed.

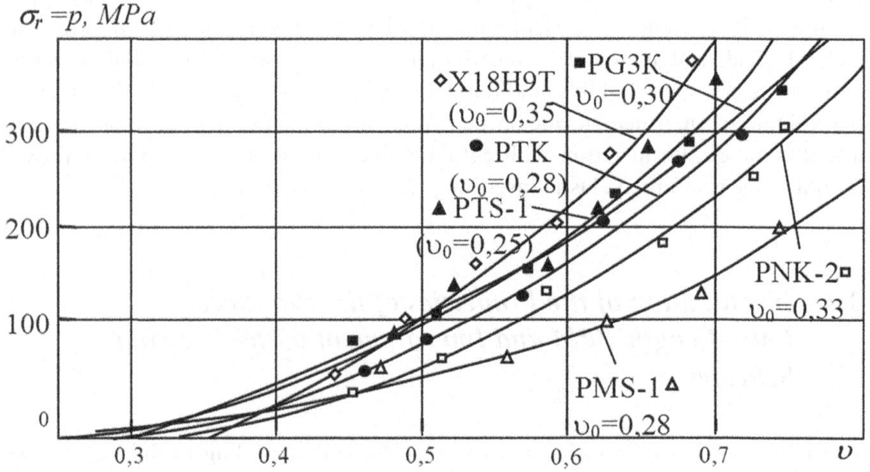

Fig. 3 Dependences of relative density on compression pressure following dry radial-isostatic pressing (DRIP) of powder permeable materials (PPM) from various metal powders. The curves are constructed according to Eq. (1). The points are the experimental data

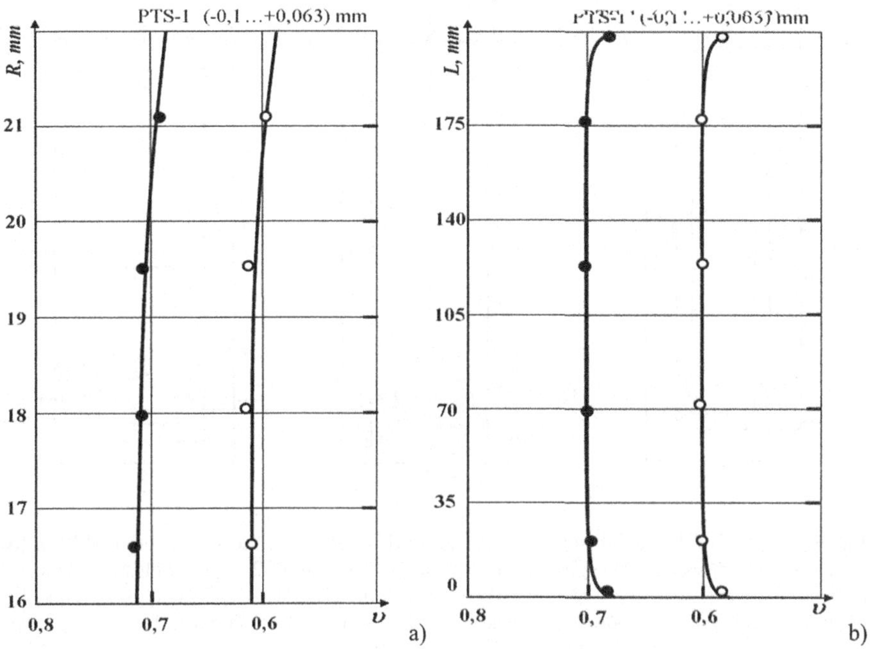

Fig. 4 Experimental dependencies of density distribution in briquettes obtained from titanium powder PTS-1 (−0.1 ... +0.063) mm: (**a**) along the radius of the powder tube with dimensions ⌀44 × ⌀32 mm; (**b**) along the length of the powder tube with dimensions ⌀44 × ⌀32 × 210 mm. The points are the experimental data

Investigation of experimental data on the density distribution along the length of tubular powdered briquettes obtained from PTS-1 titanium powder with finite dimensions ⌀44 × ⌀32 × 210 mm shows that the density of the product is uniformly distributed over the length of the product (see Fig. 4b). However, there are certain edge effects of a significant decrease in density at the ends of the briquette. In our opinion, this is due to the presence of external friction during compaction of the powder on the end surface of the mandrel. Accordingly, in these areas of reduced density, the characteristics of the porous structure differ from the design, so these areas are usually removed. This is a negative factor as it leads to the formation of a significant amount of waste of expensive powder material.

To eliminate the appearance of this negative effect, we propose to introduce to the design adaptations of the special elastic inserts (2) (see Fig. 5a), which are dressed with a slight tension on the mandrel (1) from both ends [5]. Moreover, first, the elastic insert is worn on the mandrel, after which the necessary quantity of powder (3) is used to fill up the apparatus and then the second insert is worn on top and the adaptations continue. These elastic inserts help to avoid external friction between the powder and the end face of the mandrel. In addition, in the process of radial compression, elastic inserts, due to the flow of the elastomer in the axial direction,

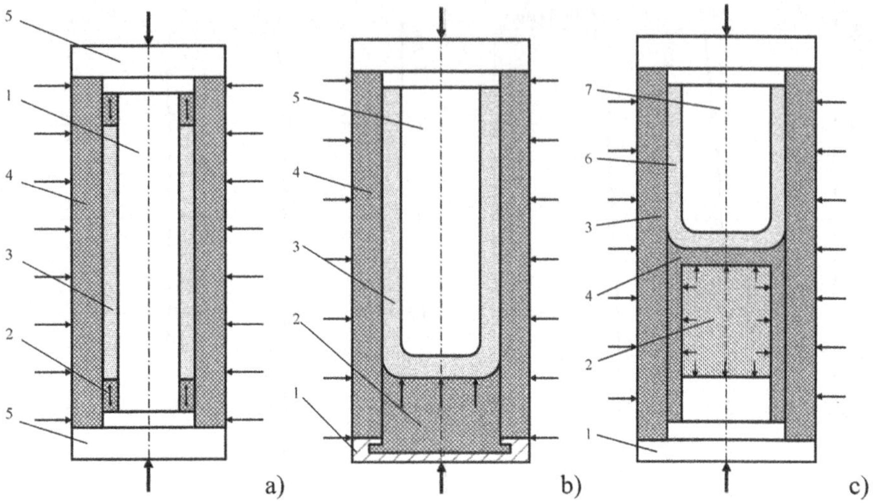

Fig. 5 Schemes of dry radial-isostatic pressing (DRIP) powder permeable materials (PPM): (**a**) tubular form with special elastic inserts: 1 – shaping rod (mandrel); 2 – special elastic inserts; 3 – powder; 4 – elastic shell; 5 – end stop; (**b**) complex form "glass" type with rigid elastic cork: 1 – plug; 2 – elastic cork; 3 – powder; 4 – elastic shell; 5 – mandrel; (**c**) complex form "glass" type with elastic cork filled with liquid: 1 – end stop; 2 – liquid; 3 – elastic shell; 4 – elastic cork filled with liquid; 6 – powder; 7 – mandrel

redistribute pressure to the end of the melding. This will allow to obtain high-quality tubular powder briquettes with uniformly distributed density along their length. After sintering, the finished product practically does not require any further processing and provides uniform performance throughout the entire working surface of the product.

In order to obtain long-length PPM based on metal powders, except for powder pipes, it is also noteworthy that there are powder products with complex forms with non-through holes of the "glass" type. Figure 5b shows the traditional scheme of pressing this class of products [1]. In this process, there are two schemes of seals: (1) radial on a cylindrical surface and (2) axial on the end face (the bottom of the product). The quality of the obtained products depends on the properties of the used elastic materials and the geometric parameters (sizes and shapes) of the elements of the form for pressing.

The use of rigid elastic cork (2) on the end face causes a substantial decrease in the density of the bottom of the product, since the radial circuit and the properties of the elastic cork do not provide the necessary axial movements in the end face. This leads to nonuniform operational properties on the surface of the product, the appearance of micro- and macro-cracks in the periphery of the bottom is observed, and, as a consequence, the destruction of the briquette at the stage of its obtaining or at the stage of sintering. This, again, leads to a significant increase in the number of deficiencies and costs of expensive powder materials. Therefore, for obtaining long-length products of increased complexity, we propose to change the pressing scheme and bring it closer to the quasi-isostatic compression scheme [5]. This can be realized by using a special elastic cork (4) hermetically filled with liquid (2) (see Fig. 5c).

Figure 6 shows the results of an experimental study of the density distribution in briquettes of a complex form of the "glass" type of the obtained DRIP process with a rigid elastic insertion (see Fig. 5b) and the obtained DRIP process with the use of an improved elastic cork filled with liquid (see Fig. 5c).

As can be seen from the results, the application of our elastic cork, hermetically filled with liquid, completely eliminates the problem of lowering the density of the bottom of the product and allows us to obtain long-length PPM of a complex form of the "glass" type with uniform distribution densities by volume. In this way, the application of the technical solutions we have developed (see Fig. 5) allows us to obtain high-quality long-length PPM of both simple and complex forms with uniform distribution densities by volume and stable operating characteristics throughout the entire working surface.

Obtaining a more uniform distribution of density is also achieved on the basis of using the DRIP method to produce PPM with a more uniform distribution of local permeability in the area of filtration. In the course of research, a method of assessing local permeability was used based on the measurement of volumetric air flow through small areas of samples, which are examined by means of a measuring head and according to the procedure outlined in [1].

Figure 7 shows the results of the study of the deviation of the coefficient of local permeability ΔK from its average value K along the length of tubular powder samples of dimensions ⌀44 × ⌀32 mm obtained by the DRIP method. As shown in the figure, advanced technology can significantly reduce the deviation of local permeability $\Delta K/K$ in comparison with traditional technology, which will ensure

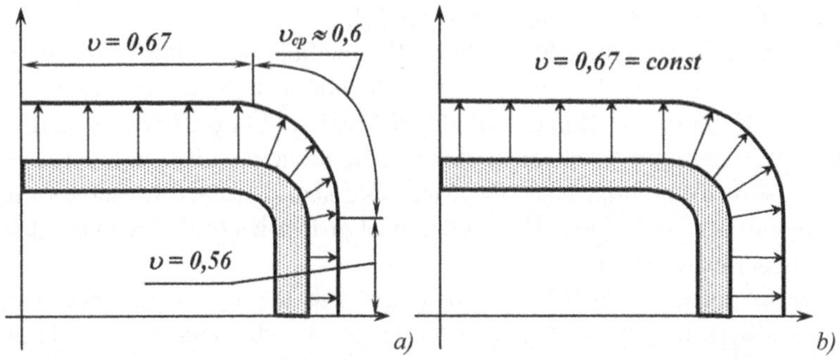

Fig. 6 Diagram of density distribution by the contour of powder permeable materials (PPM) from the titanium powder PTS-1 with particle size (−0.1 ... +0.063) mm: (**a**) using a traditional rigid elastic cork (Fig. 5b); (**b**) using an improved elastic cork filled with a liquid (Fig. 5c)

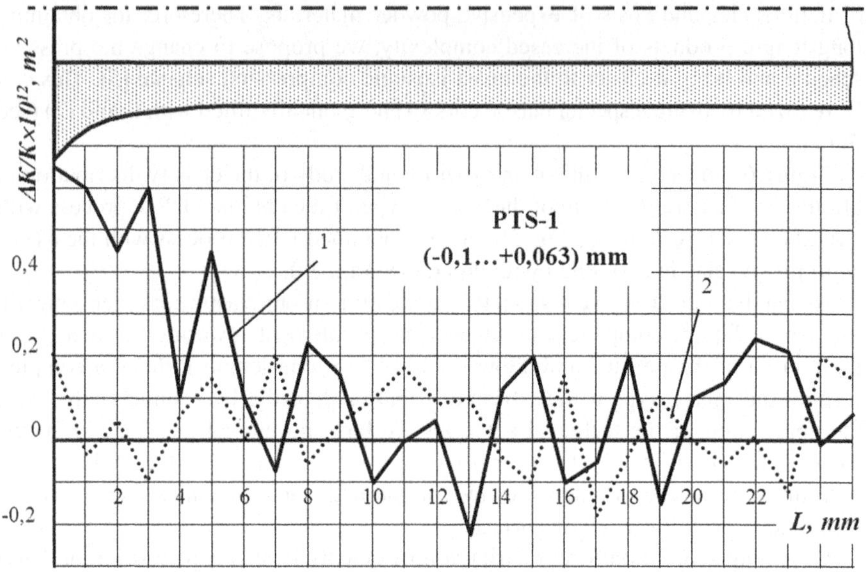

Fig. 7 Dependences of the deviation of local permeability ΔK from its average value K along the length of the powder tubular powder permeable materials (PPM) obtained by dry radial-isostatic pressing (DRIP) on the mandrel: 1 – the traditional technology without elastic inserts [1]; 2 – advanced technology with special elastic inserts (Fig. 5a)

the continuity of properties throughout the length of the product. There is no need for additional processing of the cutting of end surfaces (saving material and labor costs).

Figure 8 shows the differential curves of pore distribution by the size of the PPM with non-through axial holes of the “glass” type from PTS-1 titanium powder by the DRIP method. Differential curves of pore size distribution were determined by the method of displacement of fluid from pores [1].

Thus, in the PPM obtained by the DRIP method according to the traditional technology [1] (see Fig. 5b) (the region at the bottom), the peak of curve 2 of the pore size distribution function is shifted to the region of larger pore sizes. The DRIP method using an improved elastic cork with a hermetically filled liquid (see Fig. 5c) allows to shift differential curve 3 of the pore size distribution function into the region of smaller pore sizes, close to the level properties of the cylindrical surface of the powder material (curve 3).

In this manner, the DRIP method along with the use of our new technical solutions allows to obtain a more uniform density distribution by volume and, consequently, provides stable performance characteristics of PPM throughout their entire working surfaces.

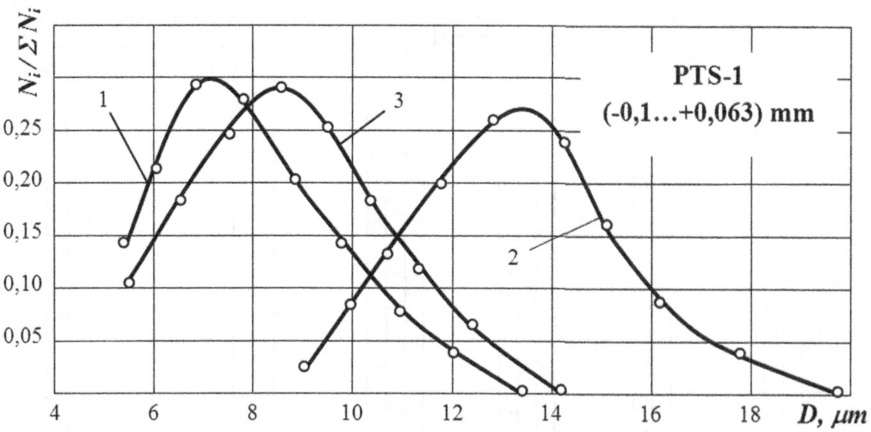

Fig. 8 Differential curves of pore distribution by sizes of powder permeable materials (PPM) of "glass" type with non-through axial holes obtained by dry radial-isostatic pressing (DRIP): 1 – cylindrical surface of "glass" type PPM; 2 – the bottom of "glass" type PPM by traditional technology (Fig. 5b); 3 – the bottom of "glass" type PPM by advanced technology (Fig. 5c)

4 Results

4.1 *Improvement of the Design of Elastic Press Tools and Fittings*

In accordance with the above-mentioned studies and proposed technical solutions for improving the pressing schemes for long-length tubular powder products, we have developed a series of adaptations (see Fig. 9a) that prevents the contact gravity of the powder in the end of the mandrel (1) and allows to create uniform pressures on the face surfaces by extruding the powder with special elastic polyurethane inserts (2) of a definite length (20–30 mm). Moreover, initially, on the mandrel (1), with the help of a special tubular pusher, the first elastic insert is worn with a small tension, after which on the vibration stand in the apparatus the necessary amount of powder (3) is used to fill it and then the second insert is worn on top and the setup continues. This allows to obtain high-quality long-length tubular PPM with uniformly distributed density throughout the length. After sintering, the finished products practically do not require processing and provide the necessary operational characteristics.

In the case of pressing tubular products of lower length than the total length of the adaptations, its construction (see Fig. 9b), when a certain portion of powder is poured into the inner cavity, is then offered, and then an intermediate elastic insert (ring) is established by a special tubular push–puller and the process is repeated until the required height is obtained. After pressing, the items are removed. The presented scheme of adaptations can increase the productivity by reducing the auxiliary time.

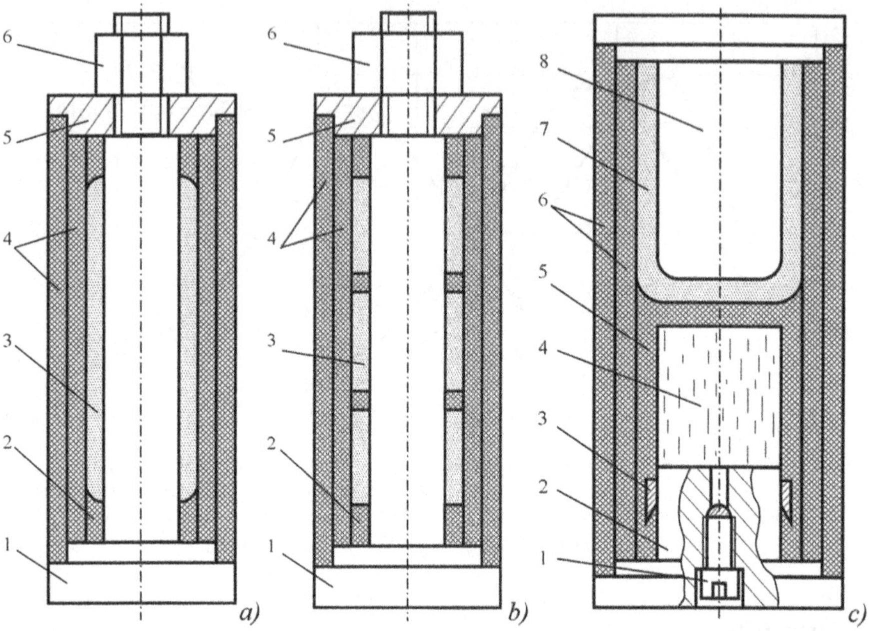

Fig. 9 Improved structures of adaptations (molds) for obtaining powder permeable materials (PPM) by the dry radial-isostatic pressing (DRIP) method: (**a**) adaptation with elastic inserts for obtaining long-length powder pipes (1 – mandrel; 2 – elastic inserts; 3 – powder; 4 – set of elastic shells; 5 – border disk; 6 – nut); (**b**) adaptation with elastic inserts for short tubular PPM (1 – mandrel; 2 – elastic inserts; 3 – powder; 4 – set of elastic shells; 5 – border disk; 6 – nut); (**c**) adaptation with elastic cork hermetically filled with liquid for obtaining "glass" type PPM with non-through axial holes (1 – screw; 2 – end stop; 3 – steel insertion; 4 – liquid; 5 – elastic cork; 6 – set of elastic shells; 7 – powder; 8 – mandrel)

For products of complex shape (glass type, ribbed surfaces), it is necessary to strive for the creation of conditions for the uniform application of pressure throughout the external contour of the powder briquette. The quality of the products obtained depends significantly on the properties of the used elastic materials and the geometric parameters (sizes and shapes) of pressing mold elements. The use of rigid elastic corks on the end face (see Fig. 5b) leads to a decrease of density in the bottom of the product, since the radial scheme and the properties of the elastic insertions do not provide the necessary axial displacements in the area of the product bottom. This leads to nonuniform operating properties on the surface. Occasionally, there is the appearance of micro- and macrocracks in the periphery of the bottom and the destruction of briquettes at the stage of pressing or sintering, which leads to a significant increase in the number of defects.

Therefore, for products of complex forms, we propose to change the compaction scheme and, approaching it, to the quasi-isostatic compression scheme using a special elastic cork (5) filled with liquid (4) (Fig. 9c). The device for DRIP of powder materials consists of a steel mandrel (8) of the corresponding shape, which is inserted into the elastic shell (6) on one side and the elastic cork (5) on the other. A compaction powder (7) is placed between the mandrel (8) and the elastic cork (5), which is filled with fluid (4) and closed with the end stop (2). The elastic cork (5) contains a metallic insertion (3) that serves to form compaction cuffs. The end stop (2) comprises an axial through hole for removing air and a screw (1) to provide hermetization of the cork (5).

The device for DRIP of PPM works as follows. The required set of elastic shells (6) are put on a steel mandrel (8), after which, in the vertical position, the required amount of powder (7) is used to fill the form. In the elastic cork (5) with the metallic insertion (3), it is filled with the liquid (4) and closed with the end stop (2). After removing the air from the elastic cork (5), the screw (1) is fitted into the stop (2) and the collected cork (5) is placed in the inner cavity of the elastic shell (6). Then, the assembled form is placed in the working cavity of the press block, where it is fixed to prevent axial displacements, and external pressure is applied. As a result, the elastic shell (6) redistributes pressure to both the cylindrical side of the forming product and the elastic cork (5), which, in turn, transfers pressure to the bottom of the product. Due to the fluid contained in the cork (as is known, a fluid uniformly transmits pressure in all directions), the pressure transmitted to the bottom of the product is practically identical to the pressure transmitted to its lateral sides. In this way, the constant working pressure across the entire product surface is ensured and, as a result, the product is formed with a uniformly distributed density by volume.

As can be seen, the use of an elastic cork, hermetically filled with a liquid, allows to obtain products of complex shape with uniform distribution density by volume. This allows to produce high-quality PPM with stable operational characteristics over their entire working surface. Thus, using the radial scheme of DRIP quasi-isostatically along with the technical solutions we have developed (see Fig. 9) will allow to obtain long-length powder products of both simple and complex forms with uniform density distributions at a wide range of pressures.

4.2 *Technology of Manufacturing Multilayer PPM*

The technology of manufacturing (two- or multilayer) PPM includes a step-by-step DRIP of powders of one composition (or a mixture of powders) on a mandrel or matrix followed by sintering of the resulting powdered briquettes in the self-propagating high-temperature synthesis (SHS) mode [6, 7] with (or without) protective atmosphere (see Fig. 10a).

In the manufacturing of two-layer PPM from powders of different fractions to create a layer with small pore sizes, we used a titanium powder PTS-1 with particle size (−0.1 ... +0.063) mm and, to create a basis layer (frame), we used a titanium powder PTK with particle size (−0.63 ... +0.4) mm.

The compressed powder briquettes were sintered in an SHS mode with a protective atmosphere and their properties were investigated: pore sizes determined according to the Ukraine state standard GOST 26849-85 and permeability coef-

Fig. 10 (**a**) Scheme of step-by-step dry radial-isostatic pressing (DRIP) titanium powder permeable materials (PPM) on a mandrel; (**b**) general view of two-layer PPM based on titanium powders: internal layer – PTK powder (−0.63… +0.4) mm; external layer – PTS powder (−0.1… +0.063) mm; (**c**) general view of different kinds of multilayer PPM (simple and complex forms)

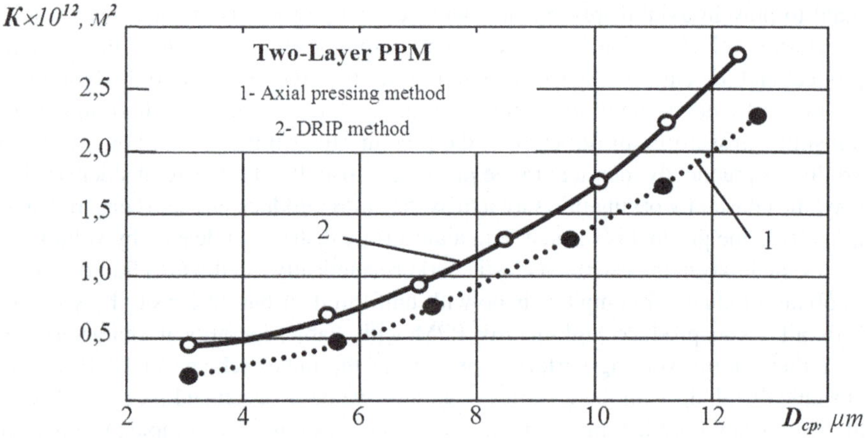

Fig. 11 Dependence of the coefficient of permeability *K* from the average pore sizes *D*: 1 – two-layer powder permeable materials (PPM, axial pressing); 2 – two-layer PPM (dry radial-isostatic pressing, DRIP)

ficient determined according to the Ukraine state standard GOST 25283-82. The general view and structure of the obtained two-layer FM based on titanium powders is shown in Fig. 10b, c. The obtained results were compared with the results given in [1], where the properties of two-layer powdered FM produced by axial pressing in a steel mold are presented.

Figure 11 shows the dependencies of the coefficient of permeability *K* on the average pore sizes *D* of PPM obtained by the DRIP method and PPM obtained by axial compression in a steel mold. The internal layer is PTK titanium powder (−0.63… +0.4) mm and the external layer is PTS titanium powder (−0.1… +0.063) mm.

The results show that use of the improved DRIP method allows to increase the permeability at a given pore size by 35–50% compared with PPM obtained by the axial pressing method. On the basis of the obtained results, experimental batches of gradient PPM of simple and complex forms were manufactured (see Fig. 10c).

The results of the tests show that using two-layer PPM (as filters) for the purification of liquid and gas environments was made possible due to their improved properties (high chemical resistance, required mechanical strength, uniform distribution of pores along the entire filtration surface, increased permeability and pollutant capacity, the possibility of regeneration), which increased the duration of operation of the obtained long-length PPM by 1.5–2.0 times. This confirms the feasibility of improved DRIP technology in the production of high-performance multilayer FM from metal and ceramic powders.

5 Conclusions

Based on the study of the density distributions of long-length PPM, we proposed new technological schemes and technical solutions for equipment, tools, and accessories for DRIP of long-length powder products of both simple and complex forms. The obtained multilayer PPM have significantly higher permeability, pollutant capacity, and service life compared to existing analogs. The developed technology of step-by-step DRIP of compaction materials allows to obtain high-performance PPM with uniform density distribution by volume, homogeneous porous structure, and stable operational characteristics.

On the basis of theoretical and experimental researches, a technology was developed for obtaining single- and multilayer PPM based on titanium powders by the DRIP method. The technology of obtaining two- or multilayer PPM based on titanium powders by the DRIP method increased the permeability at a given pore size by 35–50% compared with PPM obtained by axial pressing in a steel mold. The results of the tests show that the use of two-layer PPM based on titanium powders makes it possible, due to their improved properties, to increase the duration of operation of the filtering material by 1.5–2.0 times, which confirms the feasibility of using our proposed technology for the production of multilayer long-length PPM (FM). The developed step-by-step DRIP method of compactable materials also allows to obtain a new generation of high-performance FM with unique operational properties, which are practically impossible to obtain via other methods. This confirms the expediency of using our developed technologies for the production of powdered FM, as well as powder nozzles for abrasive jet machines [8, 9].

References

1. Reut, O., Boginskyi, L., & Petiushik, Y. (1998). *Dry isostatic pressing of compactable materials*. Minsk: Debor.
2. Ishizaki, K., Komarneni, S., & Nanko, M. (2013). *Porous materials: Process technology and applications* (p. 240). New York: Springer Science & Business Media.
3. Tumilovich, M., Pilenevich, L., & Savich, V. (2017). Porous powder materials and products based on them for the protection of human health and the environment: production, properties, applications. *Litres, 367*, 365.

4. Boginsky, L., Reut, O., Piatsiushyk, Y., et al. (2001). The development of processes of pressing of articles from powders on the bases of metals, ceramics and graphite. In *15 International Plansee Seminar* (Vol. 3, pp. 197–209). Reutte, Austria: Plansee Holding AG.
5. Zabolotnyi, O., Sychuk, V., & Somov, D. (2018). Obtaining of porous powder materials by radial pressing method. In V. Ivanov et al. (Eds.), *Advances in design, simulation and manufacturing. DSMIE 2018. Lecture Notes in Mechanical Engineering*. Cham: Springer.
6. Su, X., et al. (2014). Self-propagating high-temperature synthesis for compound thermoelectrics and new criterion for combustion processing. *Nature Communications, 5*, 4908. https://doi.org/10.1038/ncomms5908.
7. Naplocha, K. (2018). Self-propagating high-temperature synthesis (SHS) of intermetallic matrix composites. In *Intermetallic matrix composites: Properties and applications* (pp. 203–220). Elsevier. https://doi.org/10.1016/B978-0-85709-346-2.00008-X.
8. Sychuk, V., Zabolotnyi, O., & McMillan, A. (2015). Developing new design and investigating porous nozzles for abrasive jet machine. *Powder Metallurgy and Metal Ceramics, 9–10*(53), 600–605.
9. Sychuk, V., Zabolotnyi, O., & Somov, D. (2018). Technology of effective abrasive jet machining of parts surfaces. In V. Ivanov et al. (Eds.), *Advances in design, simulation and manufacturing. DSMIE 2018. Lecture Notes in Mechanical Engineering*. Cham: Springer.

Part III
Electronics and Electrical Engineering

Feasible Human Emotion Detection from Facial Thermal Images

Kimio Oguchi and Shohei Hayashi

1 Introduction

Advances in sensing technologies have yielded a variety of application fields. A typical example is sensing human status from not just physical aspects but also mental ones. The polygraph is a famous multisensor system for gathering and analyzing human responses; the prime goal is determining if the subject is being truthful by detecting changes in skin impedance, breathing patterns, and heart rate [1]. Another example is a tool to communicate with amyotrophic lateral sclerosis (ALS) patients via electromyogram (EMG) signals [2].

These are effective, but they demand that sensors be attached to the body. Therefore, in order to reduce the user's burden, noncontact and noninvasive methods are needed. One attractive solution is infrared systems; the key component, the far infrared (FIR) cameras that can gather thermal data of objects, has become small, easy to use, with lower cost. The application area of infrared systems has broadened to include medical, industrial, electrical, architectural, geosphere areas. The author's laboratory has demonstrated the detection of human respiration by thermal imaging [3].

An experiment is conducted to verify the possibility of detecting human emotion from facial thermal images captured by a commercial FIR camera. In the experi-

K. Oguchi (✉)
Graduate School of Science and Technology, Seikei University, Musashino, Japan

Faculty of Science and Technology, Seikei University, Musashino, Japan
e-mail: oguchi@ieee.org

S. Hayashi
Faculty of Science and Technology, Seikei University, Musashino, Japan

A. Beltran Jr. et al. (eds.), *World Congress on Engineering and Technology; Innovation and its Sustainability 2018*, EAI/Springer Innovations in Communication and Computing, https://doi.org/10.1007/978-3-030-20904-9_6

ment, two types of emotion, a somewhat positive feeling and a negative one, created by performing two different tasks, are distinguished.

The structure of the paper is as follows: Section 2 describes the experimental setup and procedure, and then the results and discussions are given in Sect. 3. That is followed by the conclusion and future work.

2 Experimental Setup and Procedure

2.1 Experimental Setup

Basic experimental setup is depicted in Fig. 1. Front and side face views of the subject sitting on a chair (0.45 m height) are captured by a far infrared camera (FIRC) (InfRec R300; Nippon Avionics Co., Ltd. [4]). Some details of the camera are described in the next subsection. Captured thermal images are transferred to the signal processing unit (SPU; a Laptop PC) via a USB interface.

The camera's field of view is centered on the subject's nose and covers the head-top to collarbone by adjusting the camera's height and distance from the subject. A white mattress is set 0.5 m behind the subject to provide a clear thermal background. Room temperature is held at a constant of 24–25 °C. Thermal range of the camera is fixed at 24–40 °C. The hair of the subject is kept away from the face by hairpins as necessarily needed.

Thirteen subjects participated; all were young (average age 21.1 years) healthy males (7) and females (6), as listed in Table 1. Written consent was obtained from each subject before commencing the experiment in accordance with the research ethics rules of the university.

2.2 Camera and Software Used

Major specifications of the FIRC are listed in Table 2 [4]. To analyze the captured thermal images, we used the program InfRec Analyzer NS9500 Standard (Nippon

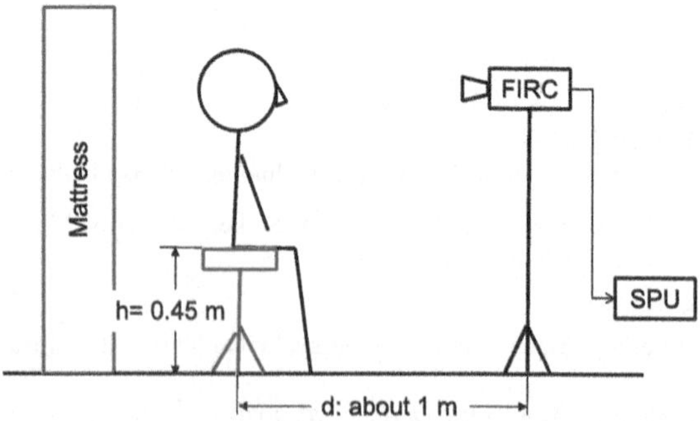

Fig. 1 Basic view of experimental setup. (SPU Signal Processing Unit, FIRC Far InfraRed Camera)

Table 1 Subjects participating

Subject	Gender	Age	Note
A	Female	22	Cold fingers (CF)
B	Female	21	CF
C	Male	21	
D	Male	21	
E	Male	21	
F	Female	21	CF
G	Male	19	
H	Male	21	
I	Male	22	CF
J	Female	21	CF
K	Female	22	
L	Female	21	CF
M	Male	21	

Table 2 Specifications of the FIRC used

Item	Specification	Note
Captured wavelengths	8–14 micrometer	
Resolution	320 (H) × 240 (V) pixel	76,800 pixels
Frame rate	1/60 sec	
Measurable temperature range	−40–500 °C	
Minimum temperature difference	0.05 °C	At 30 °C
Accuracy	±1 °C	

Table 3 Specifications of the analysis program

Item	Specification	Note
Temperature range	Max/min/middle arbitrary	
Measurable area	20 points/20 areas/line	
Peak/bottom hold	All/some areas	
Temperature variance	Point/all/ some areas	Max/min/ave
Histogram	All/ some areas	
Data output format	avi. bmp/jpg, csv/txt	

Avionics Co., Ltd. [5]). Captured images were analyzed in off-line manner. Major specifications of the program are listed in Table 3.

2.3 *Preliminary Experiment*

In order to decide which part of the face exhibited the greatest changes in temperature with emotion, preliminary experiments were performed on five subjects as they attempted to complete a jigsaw puzzle consisting of plain white pieces.

After several trials, it was found that the relative temperature change between the top of the nose and the chin was larger than any other parts, and simpler task showed clear temperature responses.

Figure 2 shows an example of the facial images captured in the preliminary experiment. Figure 2a is taken at the start of the trial and (b) after 15 min. It is clearly shown that the temperature of the nose area changed. Figure 3 shows the temporal changes in relative temperature of the top of the nose where Trel corresponds to

$$\text{Trel} = \text{the temperature of nose top} - \text{the temperature of chin} \quad (1)$$

About 1.5° change was observed during the experiment. Therefore, in this paper, temperature change of the nose is used as the index of emotion response.

2.4 Experimental Procedure

Facial thermal images of the subjects (see Fig. 4) were captured by the FIRC according to the following procedure. Experimental setup is the same as in Fig. 1, and the participating subjects are as listed in Table 1:

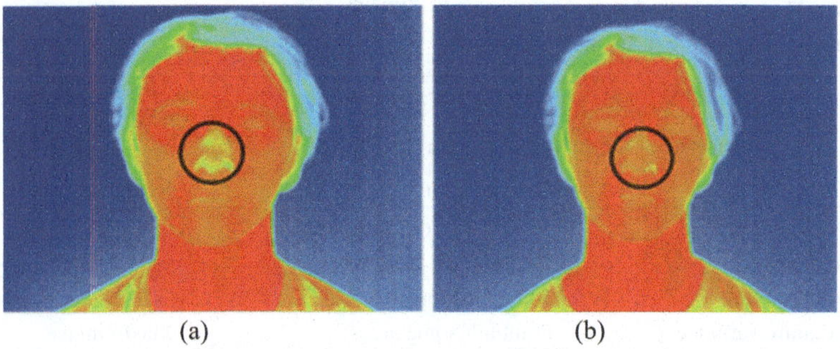

Fig. 2 Captured facial images in the preliminary experiment: (**a**) taken at the beginning of a trial and (**b**) after 15 min. It is clearly shown that the nose temperature changed

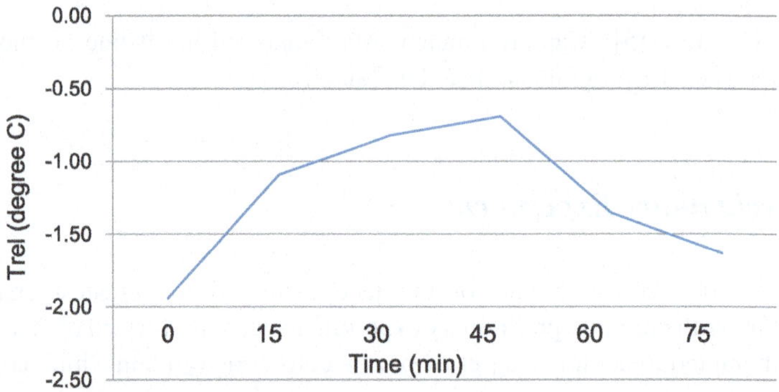

Fig. 3 Relative temperature change during trial

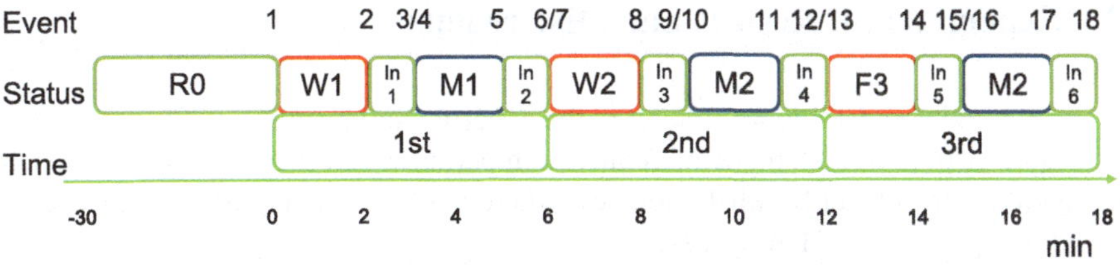

R: Rest
W: Watching movie
M: Mental maths
In: Intermission

Event	Status	Note
1, 7, 13	Start to watch movie	
2, 8, 14	End of watching movie	
3, 9, 15	1 min after the watching movie	
4, 10, 16	Start of mental maths	
5, 11, 17	End of mental maths	
6, 12, 18	1 min after the mental maths	

Fig. 4 Experimental procedure. The top pane shows time sequence of the procedure. The bottom table describes and numbers each event

Step 1: The subject sat on the chair for 30 min to achieve thermal stability.
Step 2: (Task 1) Watching a movie (detailed below) on a CRT for 2 min.
Step 3: Rest for 1 min.
Step 4: (Task 2) Engage in mental math challenge (detailed below) for 1 min.
Step 5: Rest for 1 min.
Step 6: Repeat steps 2 to 5 twice.
Step 7: Answer questionnaire described below.
Step 8: Finish.

At the start of each step, a facial thermal image was taken.

Task 1: The movie watched is titled “The world’s most relaxing film” [6]. Produced in Denmark, it is intended to reduce the heart rate by 10 + % of viewers after viewing. This is considered to yield a positive effect.

Task 2: The mental math challenge was taken from mental tests [7] intended to stress the subject. A subject vocalizes the subtraction of 2 arbitrary digits first from 4 digits, and then the subtraction is repeated (serial subtraction). If an answer is incorrect, the subtraction is repeated from the beginning.

The questionnaire required the subject to enter a score 0 (min) to 5 (max) indicating the degree of positive feeling experienced, e.g., calm or relaxed, just after the watching the movie, and a score 0 (min) to 5 (max) indicating the degree of negative feeling experienced, e.g., stress or discomfort, just after the challenge and to confirm if he/she has the cold finger effect.

After these procedures, all thermal images were analyzed to discover the temperature of the nose.

3 Experimental Results and Discussions

Figures 5 and 6 and Table 4 summarize the experimental results.

Figure 5 plots the changes of relative temperature (Trel) measured from each image for each subject for all 13 subjects where the horizontal axis plots the events listed in Fig. 4 and the Trel values.

As seen, most Trel values clearly depend on the task performed. However, three subjects (F, I, and J) showed very small Trel changes with lower nose temperature by

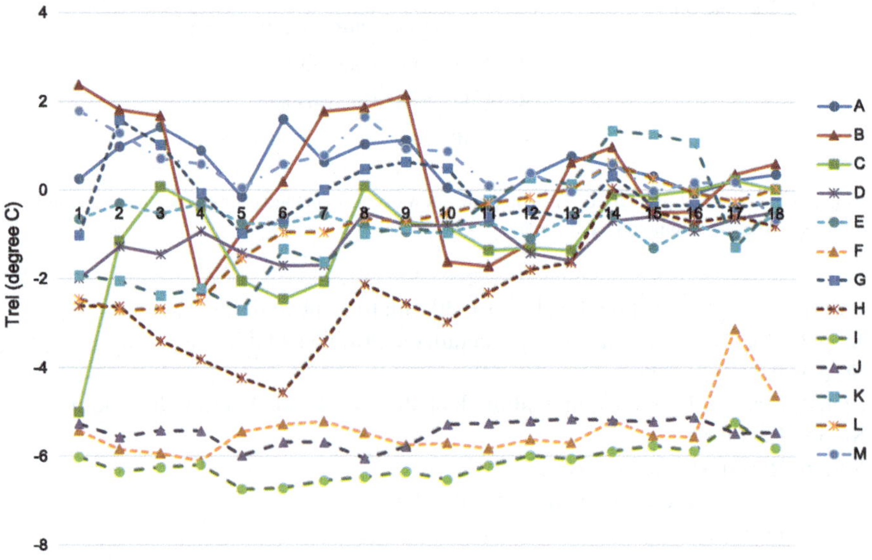

Fig. 5 Measured Trel value at each event for all subjects

Fig. 6 Relationship of positive and negative Trel values for all subjects where "x" denotes averaged value for all subjects, while "o" denotes those for subjects without the cool finger effect

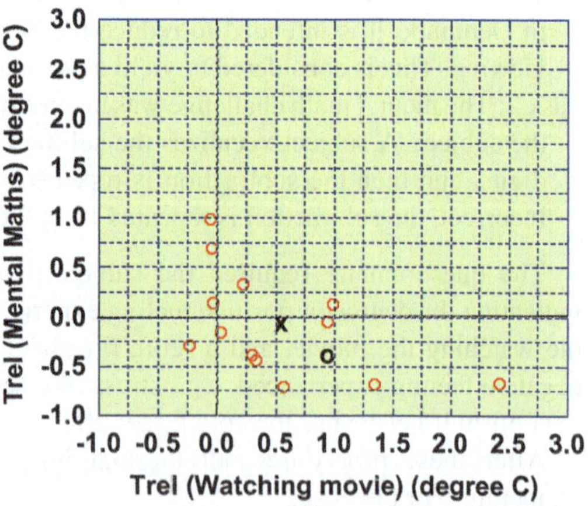

Table 4 Questionnaire results

Subject	Gender	Age	Score for WV (Task 1)	Score for MM (Task 2)	Note
A	Female	22	2	4	Cold fingers (CF)
B	Female	21	2	4	CF
C	Male	21	4	0	
D	Male	21	1	4	
E	Male	21	3	5	
F	Female	21	2	4	CF
G	Male	19	4	5	
H	Male	21	2	3	
I	Male	22	2	0	CF
J	Female	21	4	0	CF
K	Female	22	2	5	
L	Female	21	3	4	CF
M	Male	21	4	5	

about 3 °C than others. This is because all three subjects indicated that they suffered the cold finger effect [8].

Figure 6 shows positive (Task 1) and negative Trel values (Task 2) for all subjects. Measured average temperature differences were +0.52 °C for the former task and −0.08 °C for the latter task; see the mark "x" in Fig. 6. This means that Trel increases if he/she feels positive and drops if the feeling is negative. These results imply the method might be feasible. However, for subjects exhibiting the cool finger effect, the temperature variance was too small making emotion discrimination impossible. For subjects who did not exhibit the cool finger effect, measured average Trel values were +0.95 °C for the positive feeling task and −0.36 °C for the negative feeling task; see the mark "o" in Fig. 6, a difference that shows clearly the feasibility of the method.

Table 4 gives the results of the questionnaires. Scores for both tasks showed no clear difference in terms of each temperature measured.

4 Conclusion and Future Work

We conducted an experiment to verify the possibility of detecting human emotion from facial thermal images captured by an FIR camera.

In the experiment, two types of emotion, a somewhat positive feeling and a negative feeling, were created by requiring the subjects to perform different tasks. The experiment focused on the relative change in nose temperature (baseline is chin temperature). A total of 13 young healthy males/females participated in the experiment. The experiment's results showed that the relative temperature change depend on the emotion created. Measured average temperature differences were +0.52 °C for the positive feeling task and −0.08 °C for the negative feeling task.

The method is not so effective if the subject suffers from the cool finger effect; the temperature differences are too small to permit reliable discrimination of emotion type. For subjects without the cool finger effect, measured average temperature differences were +0.95 °C for the positive feeling task and −0.36 °C for the negative feeling task, which clearly show that the method is feasible.

Future work includes to verify the method with a larger number of subjects and to use multiple parts of the face.

References

1. Takazawa, N. (2009). Polygraph test; the current trends of field polygraph test and studies in Japan. *Japanese Journal of Physiological Psychology and Psychophysiology, 27*(1), 1–4.
2. Kawatsu, S., et al. (2018). Communication support for ALS patients on Minimal Communication State (MCS). *Medical Journal of Osaka General Medical Center, 40*(1), 17–19.
3. Koide, T, et al.(2009). *Breathing detection by far infrared (FIR) imaging in a home healthcare system*. ISMICT2009, Paper 3D02, Montreal, Canada.
4. Avio Homepage, http://www.avio.co.jp/english/, last accessed 2018/8/15.
5. Avio Homepage, http://www.avio.co.jp/products/infrared/lineup/software/ns9500pro.html, last accessed 2018/8/15.
6. Picles, K. The most relaxing film ever made? This seven-minute video of Danish scenery set to tranquil music is found to reduce pulse rates by more than 10%. mailOnline News, http://www.dailymail.co.uk/news/article-3092521/The-relaxing-film-seven-minute-video-Danish-scenery-set-tranquil-music-reduce-pulse-rates-10.html, last accessed 2018/8/15.
7. Yajima, J. (2012). A manual of stress experiment using mental stress testing. *Report of Beppu University, 14*, 101–107. (in Japanese).
8. Ishikawa, K., et al. (2015). *Visual methods to assess cold fingers and experimental verification*. 37th Annual International Conference of the IEEE Engineering in Medicine and Biology Society (EMBC2015), 6776–6779, Milano, Italy.

Determination of Calcium and pH Level in Urine for Calcium-Based Kidney Stone Diagnosis Using Arduino Microcontroller

Rolando Pula and Ramon Garcia

1 Introduction

Urinary tract disease and other related diseases such as kidney stone are common illnesses acquired generally by the most part of the population. The number of people having these kinds of diseases increased steadily for the past 100 years all over the world. Urolithiasis, the formation of stony concretions in the bladder or urinary tract, is considered to be one of the major clinical and economic burdens of healthcare systems around the world [1, 2]. Urolithiasis occurred when the equilibrium of urine component is disturbed and when the main marker of urolithiasis is separated from promoters and inhibitors. Increased concentrations of promoters (calcium, ammonium, phosphate, urate) indicate a risk of possible stone formation [3].The statistic shows that 12% of the world population experience renal stone disease with 70–80% in male and 47–60% in female [4]. In fact, it is one of the major causes of death all over the world [5–7]. The methods of diagnosing this kind of disease for the past years concentrate mostly on processing the image gathered through

This work was supported financially by the Philippines' Department of Science and Technology–Engineering Research and Development for Technology (DOST–ERDT) through the scholarship program for graduate students.

R. Pula (✉)
Graduate Studies, Mapua University, Manila, Philippines

School of Graduate Studies, Mapua University, Manila, Philippines
e-mail: pularolando@gmail.com

R. Garcia
School of EECE, Mapua University, Manila, Philippines
e-mail: rggarcia747@yahoo.com

A. Beltran Jr. et al. (eds.), *World Congress on Engineering and Technology; Innovation and its Sustainability 2018*, EAI/Springer Innovations in Communication and Computing, https://doi.org/10.1007/978-3-030-20904-9_7

ultrasound and other frequency-related technology. Optical crystallography, x-ray diffraction, infrared spectroscopy, x-ray spectroscopy, and thermogravimetry are some of these technologies. However, these techniques require elaborate apparatus and are generally semiquantitative and do not detect minor constituents of mixed calculi [8].

To get complete information about the chemical composition and physicochemical principles, there is a need for more precise information, which involves chemical quantitative analysis. It is being done by studying and analysing the urine sample composition. For many years, urine analysis plays an important role in diagnosing different kinds of diseases. The composition of urine which is positive of having kidney stone particles comprised mostly of calcium and calcium compounds. These tiny particles in the form of crystals will travel through the urinary tract and excreted out to the body [9]. Moreover, the pH level of urine also changes based on its basicity and acidity caused by its chemical composition and types of stone present [10].

From the above information, the possibility of creating a system for possible risk factors of kidney stone diagnosis based on calcium and pH level is great. This study will focus on analysing the calcium and pH level of urine using a microcontroller. Many studies' findings suggested that the level of calcium and calcium compounds contributes a great percentage in urine composition [10]. Calcium and calcium compounds found in human urine (stone former) comprised 70–80 percent, and the most dominant calcium compounds are calcium oxalate and calcium phosphate. Another property that could indicate the presence of kidney stone is the power of hydrogen (pH) level of the urine sample(s) [11]. Patient's urine having a pH level from 5.8 to 6.2 suggests the presence of calcium phosphate (renal tubular acidosis).

2 Materials and Design Implementation

Calcium sensor and pH sensor were interfaced to microcontroller after signal conditioning circuit stage. These sensors will respond to a solution tested inside a container with the sample. A separate signal conditioning circuit was created for pH and calcium sensor; this is to clean and polish first the signal before entering the Arduino microcontroller. TLV431 voltage reference IC was used for pH sensor to correct the output voltage requirements. Unity gain amplifier IC which is the LMP7721 was also used because of its low bias current that could not affect the pH sensor's function. Since the calcium sensor has built-in amplifier, unity gain circuit was the only circuit used. Signal from the conditioning circuit of pH and calcium sensors was connected to pin A2 and A3. Arduino microcontroller board was used to interpret the values gathered from the sensors (calcium and pH). The built-in A/D converter was responsible for interpreting those values from analog to digital. The system's block diagram is illustrated in Fig. 3. The exploded diagram of the final prototype can be seen in Fig. 1.

Signal conditioning circuit used in the system was shown in Fig. 2. The interconnection of the electronic devices was showed using a label in each pin (Fig. 3).

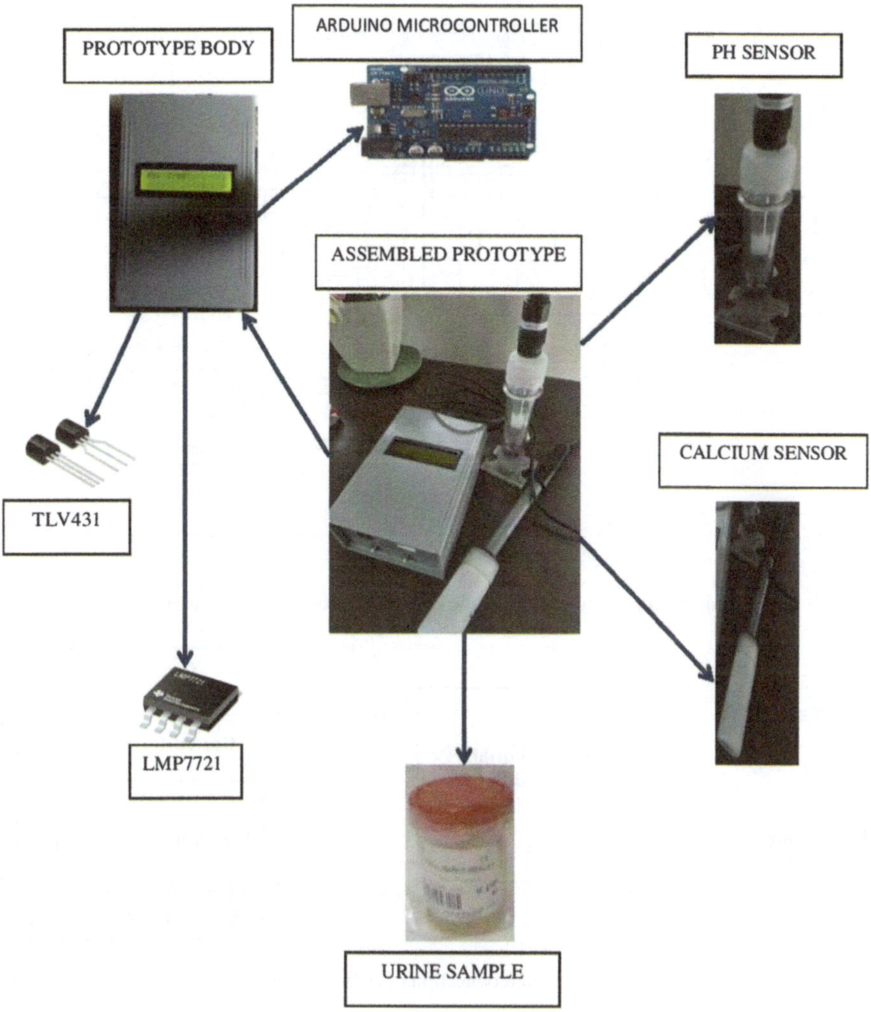

Fig. 1 Exploded diagram of the final prototype

3 Software Development

C++ language was used in coding the program of the prototype. C++ was also used to interface and process the input of calcium and pH sensors to microcontroller and finally display the result of those sensors. C++ is a widely used programming language that can be used mostly by the microcontroller because of its availability and ease of access. Figure 4a, b show the sequence of the program. Each sensor has its individual sets of flow that need to be followed to be able to get the required reading.

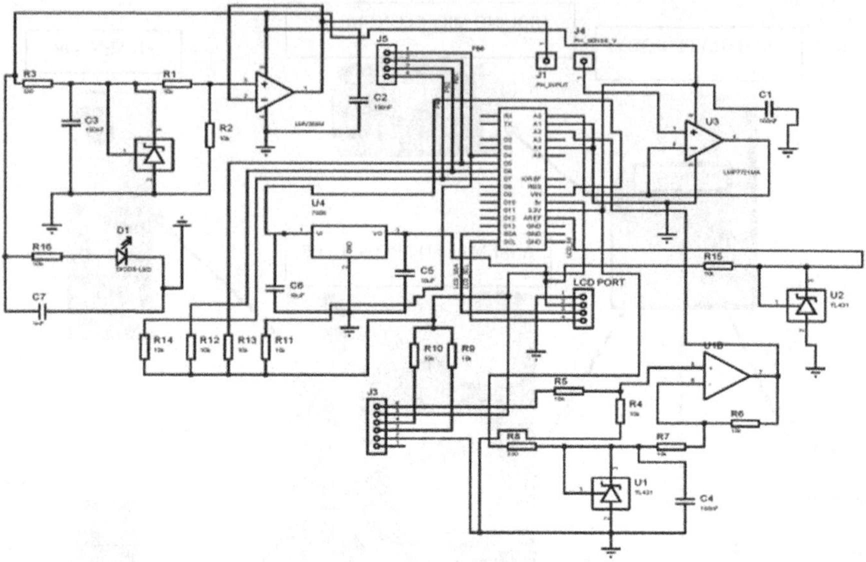

Fig. 2 Signal conditioning circuit showing the interconnection of each component

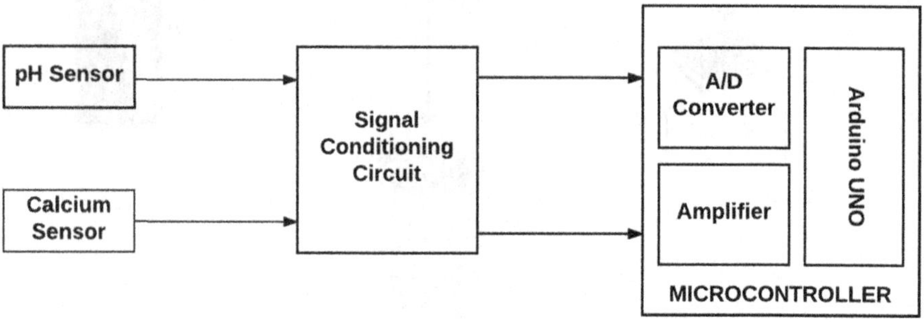

Fig. 3 Block diagram showing the interconnection of sensors to microcontroller

4 Calibrations and Testing

Values from the initial reading of the sensors have deviations observed. These deviations are normal due to the noise produced by the circuit and the degradation of the chemicals inside the sensors. Initial reading for the pH level has a minimal deviation, while the calcium sensor has a deviation value of approximately 0.4 volts in the final reading. These deviations were corrected by analysing and correcting the lower and upper limit of voltage response of each sensor in calibration process. Voltage response of calcium sensor has 1.9 V (1000 mg/L), 1.7 V in 100 mg/L, and

1.5 V (10 mg/L) based on the specification, but the measured value was ranging from 1.1volts to 1.5 volts. By using the equation that converts mg/L to mV which is

$$\text{Calcium (mV)} = -11.96^*\left(-14.942 - \ln\left(\text{"Calcium (mg/L)"}\right)\right) - 168.23 \tag{1}$$

and substituting, the value gets Calcium (mV) in the final formula which is

$$V = 0.00727^*\text{Calcium (mV)} + 1.223 \tag{2}$$

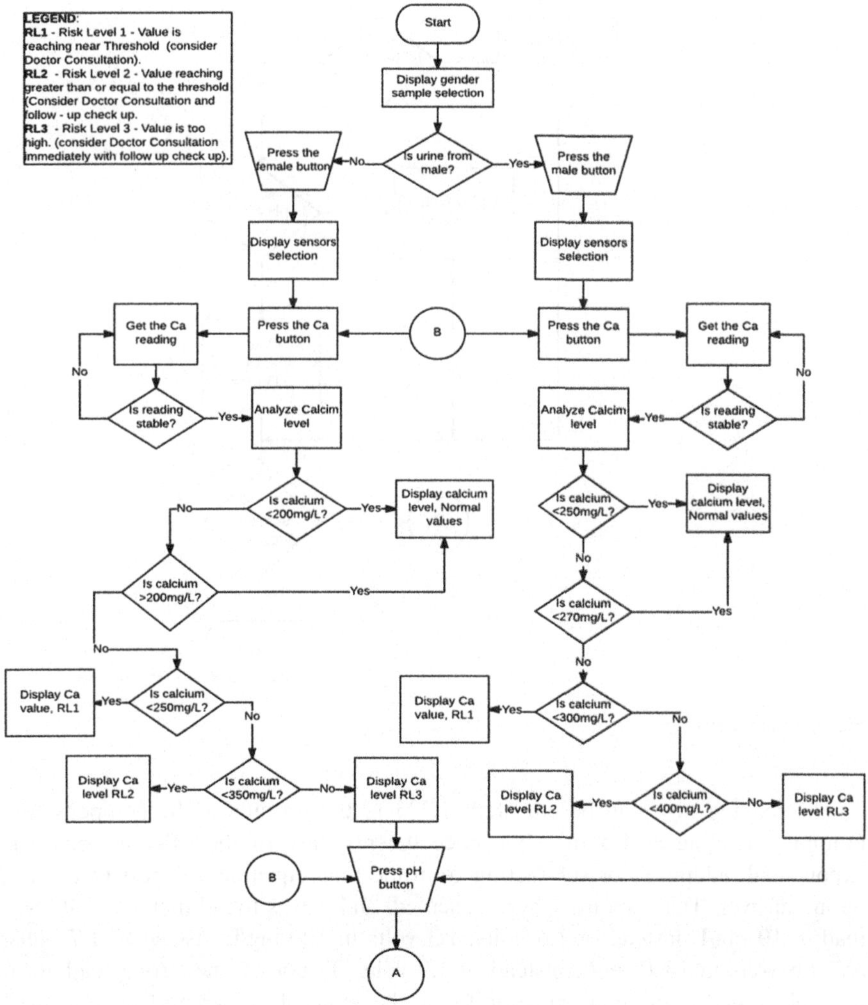

Fig. 4 (**a, b**) System program flowchart

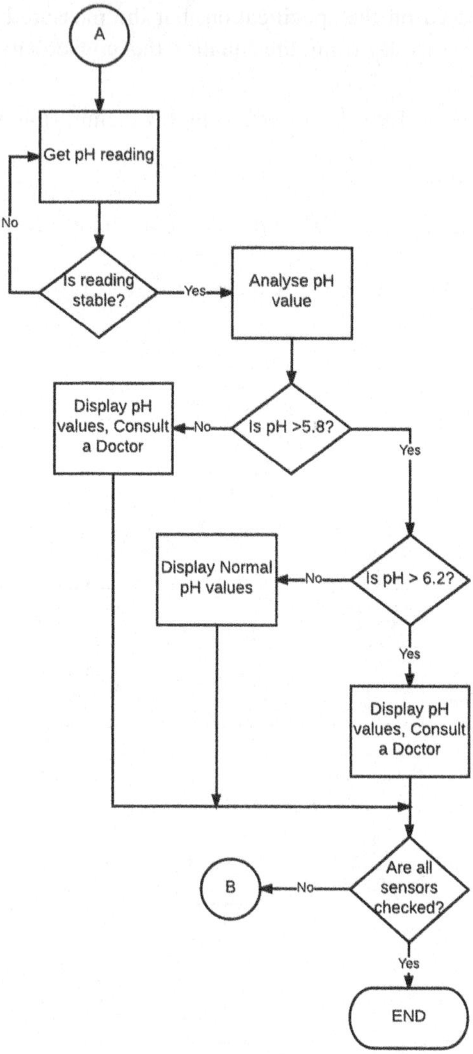

Fig. 4 (continued)

The reading now can be adjusted; 1.223 in the formula is the compensation minimum reading at 1.5 mV, but since the response of the calcium sensor is exponential, adding to or subtracting from the final equation will just result in a wrong answer. The responses were observed, and it was found that 1.1 volts was read in 10 mg/L instead of 1.5 volts, 1.3 volts in 100 mg/L instead of 1.7 volts, and 1.5 volts in 1000 mg/L instead of 1.9 volts. To correct the wrong reading, it was substituted in the final equation to get the value of the wrong Calcium (mV) and compared it to the correct reading of Calcium (mV). It was found that when the wrong reading of Calcium (mV) is subtracted from the correct reading, there is a common difference of 53.64511692 mV in 10 mg/L, 100 mg/L, and 1000 mg/L; these are the equivalent of 0.4 volts deviation in the final reading. To correct the reading, 53.64511692 were added to Calcium (mV) in the final equation in getting the voltage response. The new equation used in the program was now:

$$V = 0.00727^{*}(\text{Calcium (mV)} + 53.64511692) + 1.223 \tag{3}$$

In the case of pH sensor's voltage response, it ranges from −0.5 volts to 0.5 volts from the datasheet of the pH sensor. Circuit adjustment was made to avoid negative reading of pH. To do this, circuit was created to bring up the range to positive values by using voltage divider and LMP7721; the new range is from 0.1 volts to 1.1 volts. By analysing the voltage response per pH in buffer solutions and using ratio and proportion in the program, the new range of voltage can now react to pH in the solution. The measurement taken in using the buffer solution was calibrated using this principle.

Five urine samples were analysed based on a 24-hour urine collection which is the standard procedure for getting samples in kidney stone diagnosis. One female and four male subjects were tested following the proper guidelines in collecting urine without altering their diets and daily routine. Subjects were advised to collect their urine in the morning after they woke up. First urine discharge in the morning after waking up is not included in the samples, but this is the starting sign to start the time of counting based on 24 h. The second urine discharge was the first urine to be collected and put into the container. Urine after waking up the next day was included. The subjects were also advised that 10 min before and after the set time of start of collection is the buffer time. Meaning all urine 10 min before the start time of the previous day should be collected 10 min after the buffer was the stop of the collection. During the collection, subjects were advised to collect all urine discharge. They were also advised to do their normal routine and diet on the day of collection. Subjects were also advised that the collected urine samples during the collection period should be stored in a cool, dry, and dark place. This is to avoid the effect of light and temperature to the solution. Collected urine from five different subjects can be seen in Fig. 5; the total urine volume collected are in the upper portion of the container, while in the lower portion are the separate samples used for testing.

The total volume of urine and calcium and pH level were the parameters being measured in the laboratory, but only the level of pH and calcium were being tested in the designed prototype. Separate samples were requested after measuring the total volume of each sample. Requesting for a separate measurement of volume was needed since the volume of urine excretion is based on a 24-hour urine collection was also a parameter being observed to know the possibility of having kidney stones. The requested samples are 60 mL each. Three trials were performed to get the calcium and pH level of each sample. The pH level was the first parameter being

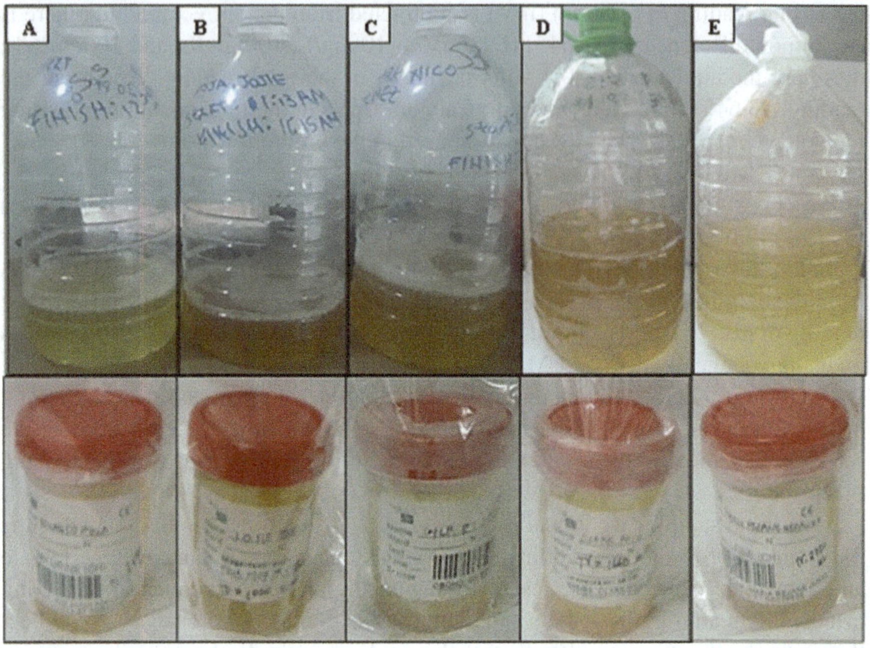

Fig. 5 Urine samples from five different subjects. Samples in the lower portion of a small container were the requested samples from the laboratory after measuring the volume. *Male subjects*: A, B, C, and D. *Female subject*: E

Table 1 Laboratory results from Hi-Precision Diagnostic Company and prototype testing results

Laboratory results						Prototype testing results	
Subjects	Gender	Age (years)	24 h urine vol. (mL)	pH	Calcium (mg/L in 24 h)	pH	Calcium (mg/L in 24 h)
A	Male	30	2580	6	136.4	5.96	136.65
B	Male	21	1000	6.1	24.8	6.08	24.60
C	Male	30	1250	6.6	48	6.54	48.11
D	Male	23	1660	6.36	132.8	6.26	132.43
E	Female	23	2720	6.5	53.2	6.56	53.245

measured followed by the calcium level. After the first measurement, the probes were cleaned to perform another trial. The procedure was repeated until all samples were tested. After testing, the mean values of the three trials in each sample were recorded. Proper guidelines in using the sensor probe were followed to get more accurate results. Laboratory analysis of the urine brought to the diagnostic centre was compared to the prototype testing results. Tables 1 and 2 show the mean value from three trials of each sample compared to the laboratory results.

Table 2 Prototype testing results

Test subjects	Parameters (Ca in mg/L for 24 h)	Trial results			Mean values
		Trial 1	Trial 2	Trial 3	
A	Calcium level	135.73	136.09	137.2	136.645
	pH level	6.21	5.76	5.9	5.956
B	Calcium level	25.15	24.45	24.21	24.603
	pH level	5.9	6.15	6.2	6.083
C	Calcium level	48.32	47.89	48.13	48.113
	pH level	6.7	6.02	6.9	6.54
D	Calcium level	132.32	133.01	131.95	132.427
	pH level	6.32	6.37	6.09	6.26
E	Calcium level	53.42	53.21	53.11	53.247
	pH level	6.57	6.45	6.67	6.563

Table 3 Descriptive statistical results of urine sample observed and laboratory tested

Parameters tested	Mean		Std. deviation
	Statistic	Std. error	Statistic
Calcium observed mg/L in 24 h	79.040	23.188	51.849
Calcium laboratory mg/L in 24 h	79.007	23.188	51.850
pH observed	6.312	0.115	0.256
pH laboratory	6.280	0.121	0.270

5 Results and Discussion

Data after testing were analysed using the statistical analysis mentioned above. Calcium and pH readings were analysed individually. Mean values of three trials in each sample were compared with laboratory result.

Urine samples obtained had undergone laboratory testing for pH level and calcium level confirmation. Each sample was requested to be verified and run for three trials. Observed samples were also tested for three trials each on five subjects. Mean values, standard error, and standard deviation were acquired for the two parameters (Table 3). Calcium observed has a statistical mean of 79.040 mg/L in 24 h. While in laboratory, results have 79.007 mg/L in 24 h having the same standard errors of 23.188 and standard deviation of 51.849 and 51.850, respectively. Mean statistical value of pH level that was observed is 6.312 closed to pH laboratory results which was 6.280, standard errors are 0.115 and 0.121, and standard deviations are 0.256 and 0.270, accordingly. Descriptive statistics was followed by paired t-test result to confirm statistical mean differences of the two parameters.

The null hypothesis of the study by using paired t-test indicates that there are no significant differences in the mean values of calcium level prototype tested versus calcium level laboratory tested and pH level prototype tested versus pH level laboratory tested. Conversely, alternative hypothesis indicates that there is a significant difference between the two paired parameters (Ho, $\mu 1 = \mu 2$; Ha, $\mu 1 \neq \mu 2$).

Based on the p values of calcium level, prototype tested versus calcium level laboratory tested which is 0.781 and pH level prototype tested versus pH level laboratory tested which is 0.310 indicated that the null hypothesis is accepted. Therefore, results from paired t-test showed that there is no significant difference in the mean values of calcium level that was observed and laboratory tested ($p = 0.781$). Accordingly, standard deviation values and standard error are 0.249 and 0.111 with a minimum effect on the samples. Mean paired difference of pH level that was observed and laboratory tested is equal to 0.036 which also showed no significant differences on values tested ($p = 0.310$). Standard deviation is equal to 0.061, and standard error is 0.027. This shows that the values obtained from two different methods such as using sensor versus laboratory testing imply closely related values provided by almost 0 mean differences of the two parameters used (Table 4).

The comparison of the laboratory results of each sample from the mean between three trials of values from testing was shown in Fig. 6 which was the pH reading and Fig. 7 which was the calcium reading. As observed from Fig. 6, the graph's critical deviation was from subject number four, and minimum deviation can be observed from subject number two. Slight changes can be observed from the laboratory result of calcium compared with the testing results as shown in Fig. 7.

Table 4 Statistical analysis result of observed and laboratory-tested calcium level and pH level using paired t-test

		Paired differences			
		Mean	Std. deviation	Std. error mean	Sig. (two-tailed)
Pair 1	Calcium observed – calcium laboratory	0.033	0.248	0.111	0.781
Pair 2	pH observed – pH laboratory	0.032	0.061	0.027	0.310

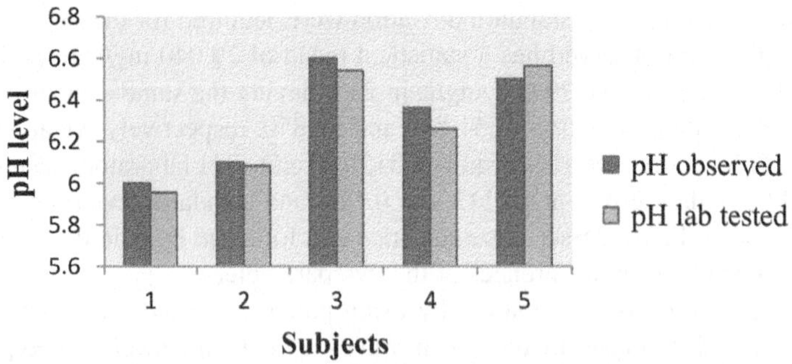

Fig. 6 Graph showing the results from laboratory and testing of pH

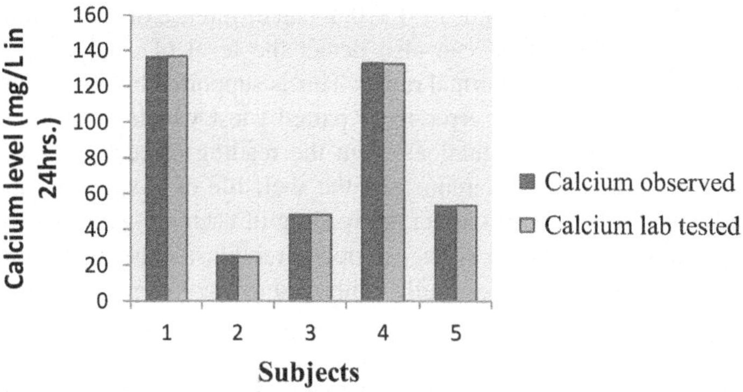

Fig. 7 Graph showing the results from laboratory and testing of calcium

Table 5 Statistical analysis result of calcium concentration compared with mean values observed from the prototype using paired t-test

		Paired differences			Sig. (2-tailed)
		Mean	Std. deviation	Std. error mean	
Pair 1	Calcium concentration – mean values observed	0.169	1.646	0.823	0.850

Since calcium kidney stone was not obtained from the samples, calcium concentrations were formulated higher than the normal ranges for males and females such as 350 mg/L, 400 mg/L, 500 mg/L, and 750 mg/L for positive controls. The null hypothesis from the statistical test performed was that there are no significant differences from the mean values of calcium concentration compared with mean values observed from the prototype. Therefore, the null hypothesis was accepted because the p value is equal to 0.850. The mean difference of the paired data set is 0.169 with a standard deviation of 1.646 and a mean standard error of 0.823.

Based on related literatures and urologist opinion, values of the calcium measured from five subjects were negative from having hypercalciuria since female subject's urine calcium level did not exceed to 250 mg/L in 1 day; the same condition was observed for male subjects. All male subjects' urine did not exceed to 300 mg/L in 1 day. In terms of pH level, three subjects exceed the basic limit which is pH 6.2. They are subjects A (pH 6.54), B (pH 6.26), and C (pH 6.56), and based on the study of [12], this indicates that calcium oxalate stones are forming in these subjects (Table 5).

6 Conclusion

In urine samples comparing with the laboratory results, the prototype can detect calcium level with a p value of 0.781, mean difference of 0.033, and mean standard error of 0.111 between laboratory and prototype testing results. The prototype can also detect pH level with a p value of 0.310, mean difference of 0.032, and standard error of 0.027. The prototype can also detect the level of calcium greater than 300 mg/L which exceeds the normal range. This is supported by the results of mean difference and mean standard error from paired t-test which is 0.169 and 0.823 with a p value of 0.850. Fluctuations from the reading of each sensor were due to the noise generated by the circuits and the shelf life of the sensors used. Some compounds found in urine also affect the reading of each sensor. Negative reading of pH sensor was avoided using voltage divider circuit inside the signal conditioning stage. Calcium sensor was successfully calibrated to the system with no significant difference.

Acknowledgement This study is funded by the Department of Science and Technology (DOST), under the scholarship program for master's degree in Engineering Research and Development for Technology through the academic support of Mapúa University.

References

1. Faridi, P., Seradj, H., Mohammadi-Samani, S., Vossoughi, M., Mohagheghzadeh, A., & Roozbeh, J. (2014). Randomized and double-blinded clinical trial of the safety and calcium kidney stone dissolving efficacy of Lapis judaicus. *Journal of Ethnopharmacology, 156*, 82–87.
2. Zarasvandi, A., Heidari, M., Sadeghi, M., & Mousapoor, E. (2013). Major and trace element composition of urinary stones, Khuzestan province, southwest, Iran. *Journal of Geochemical Exploration, 131*, 52–58.
3. Yaroshenko, I., Kirsanov, D., Kartsova, L., Sidorova, A., Borisova, I., & Legin, A. (2015). Determination of urine ionic composition with potentiometric multisensor system. *Talanta, 131*, 556–561.
4. Tiwari, A., Nipate, S., Bandawane, D., Bhandarkar, A., Londhe, V., & Soni, V. (2012). An overview on potent indigenous herbs for urinary tract infirmity : Urolithiasis. *Asian Journal of Pharmaceutical and Clinical Research, 5*(0974–2441), 7–12.
5. Jabbar, F., Asif, M., Dutani, H., Hussain, A., Malik, A., Kamal, M., & Rasool, M. (2015). Assessment of the role of general, biochemical and family history characteristics in kidney stone formation. *Saudi Journal of Biological Sciences, 22*(1), 65–68.
6. Curhan, G., & Taylor, E. (2008). 24-h uric acid excretion and the risk of kidney stones. *Kidney International, 73*(4), 489–496.
7. Taylor, E., Stampfer, M., & Curhan, G. (2005). Obesity, weight gain, and the risk of kidney stones. *JAMA, 293*(4), 455.
8. Jawalekar, S., Surve, V., & Bhutey, A. (2010). The composition and quantitative analysis of urinary calculi in patients with renal calculi. *Nepal Medical College Journal, 12*(3), 145–148.
9. Alok, S., Jain, S., Verma, A., Kumar, M., & Sabharwal, M. (2013). Pathophysiology of kidney, gallbladder and urinary stones treatment with herbal and allopathic medicine: A review. *Asian Pacific Journal of Tropical Disease, 3*(6), 496–504.
10. Laube, N., Pullmann, M., Hergarten, S., Schmidt, M., & Hesse, A. (2003). The alteration of urine composition due to stone material present in the urinary tract. *European Urology, 44*(5), 595–599.
11. Robertson, W. (2012). Methods for diagnosing the risk factors of stone formation. *Arab Journal of Urology, 10*(3), 250–257.
12. Worcester, E., & Coe, F. (2010). Calcium kidney stones. *New England Journal of Medicine, 363*(10), 954–963.

Servo-Controlled 5-Axis 3D Printer from an Open-Source Kit

Dawn Christine P. Corpuz, Ramon Miguel Imbao, Carlos M. Oppus, and Juan Antonio G. Mariñas

1 Introduction

CNC machining, or any type of subtractive manufacturing, allows users to copy objects through digital design (e.g., CAD) and turn them into physical objects through subtraction of materials via milling or cutting, etc. Objects created in the CAD tools are built as representations that are created by the user [1].

3D printing also gives the same features as subtractive manufacturing in terms of its infinite customizability and the ability to create objects from digital designs. However, the cost of materials is lower for 3D printers compared to a CNC machine because of its almost zero waste of material [2].

3D printing is recently becoming more and more readily available to home consumers, giving them the ability to create their own items easily and quickly, compared to before, where commercial 3D printer was only available in industrial settings. The availability of home 3D printers is thanks to the RepRap project which began in 2004. RepRap is short for replicating rapid prototyper, where they are 3D printers that can create copies of themselves for easier "[accessibility for] small communities in the development world as well as individuals in the developed world" [3]. The main method of 3D printing used by RepRap machines is the fused deposition modeling (FDM) method. FDM is the method of depositing molten filament layer-by-layer from the bottommost layer going upward in order to form the 3D object being created.

D. C. P. Corpuz (✉) · R. M. Imbao · C. M. Oppus · J. A. G. Mariñas
Department of Electronics, Computer and Communications Engineering, School of Science and Engineering, Ateneo de Manila University, Quezon City, Philippines
e-mail: corpuzdawn@gmail.com; ramonimbao@gmail.com; coppus@ateneo.edu; jmarinas@ateneo.edu

A. Beltran Jr. et al. (eds.), *World Congress on Engineering and Technology; Innovation and its Sustainability 2018*, EAI/Springer Innovations in Communication and Computing, https://doi.org/10.1007/978-3-030-20904-9_8

Fig. 1 5-axis_3D_printer

However, FDM has limitations. It is not able to print overhanging parts without support structures. For 3D-printed objects, overhanging areas usually mean any areas being printed that has an angle less than 45° with respect to the build platform. The solution for these overhangs, in subtractive manufacturing, is through the use of additional rotational axes like a 5-axis machine which consists of the three-dimensional Cartesian coordinate axes (x, y, and z) along with two rotational axes.

To solve the problem of FDM machines, we propose a 5-axis 3D printer machine (Fig. 1) consisting of servomotors for the rotational axes.

2 Objectives

The aim of this paper is to reduce material usage and print time from printing support structures by trying to eliminate the need for them. This can be done by first assembling and then modifying a 3-axis 3D printer to a 5-axis 3D printer using a servo pan/tilt system. Then, developing of a g-code parser to take existing g-code generated by existing 3D slicers, which are programs that take 3D objects and slice them into the individual layers to be printed, should be done next and then finally, making it compatible with the proposed 5-axis system.

3 Theoretical Framework

The 5-axis system done in this research is based on a classification of 5-axis machines by Makhanov and Anotaipaiboon [4]. The classification is based on the number of axes carried by the table and by the tool. The system is a 1/4 machine where one axis is carried by the table and four (two rotational and two translational) axes are carried by the tool. There is an existing 5-axis 3D printer research by Oyvind Grutle. For his thesis, he uses a 3/2 machine [5].

The addition of the rotational axes to the printer requires some post-processing of the g-code since the default output is only for 3 axes. For our thesis, the x, y, and z coordinates are extracted from the g-code and placed inside a $N \times 3$ matrix. These coordinates are expressed in the following matrix definition. These equations were used to multiply to the coordinate matrix to apply a rotation on a given axis [6]:

$$\boldsymbol{R}_x(\theta) = \begin{bmatrix} 1 & 0 & 0 \\ 0 & \cos\theta & \sin\theta \\ 0 & -\sin\theta & \cos\theta \end{bmatrix} \tag{1}$$

$$\boldsymbol{R}_y(\theta) = \begin{bmatrix} \cos\theta & 0 & -\sin\theta \\ 0 & 1 & 0 \\ \sin\theta & 0 & \cos\theta \end{bmatrix} \tag{2}$$

$$\boldsymbol{R}_z(\theta) = \begin{bmatrix} \cos\theta & \sin\theta & 0 \\ -\sin\theta & \cos\theta & 0 \\ 0 & 0 & 1 \end{bmatrix} \tag{3}$$

After transforming the matrix, the coordinates are now located in a position that's different relative to the nozzle at the default servo angles. Thus there is a need to compensate for the new position. The new position is a function of the servo angles. By analyzing the geometry of the servo arm assembly and using trigonometry to extract the x and y components of the servo angles, the following are the formulas determining the nozzle compensation:

$$X_{NC}(\theta_{S0}, \theta_{S1}) = \frac{X_{\text{off}}}{2} + \cos\left(\theta_{S0} - \arctan\left(\frac{X_{\text{off}}}{Y_{\text{off}}}\right)\right) \times \sqrt{\left(\frac{X_{\text{off}}}{2}\right)^2 + \left(\frac{Y_{\text{off}}}{2}\right)^2} + D\sin(\theta_{S1})\sin(\theta_{S0} + 90^\circ) \tag{4}$$

$$Y_{NC}(\theta_{S0}, \theta_{S1}) = \frac{Y_{\text{off}}}{2} + \cos\left(\theta_{S0} - \arctan\left(\frac{X_{\text{off}}}{Y_{\text{off}}}\right)\right) \times \sqrt{\left(X_{NO} - \frac{X_{\text{off}}}{2}\right)^2 + \left(Y_{NO} - \frac{Y_{\text{off}}}{2}\right)^2} + D\sin(\theta_{S1})\cos(\theta_{S0} + 90^\circ) \tag{5}$$

$$Z_{NC}(\theta_{S0}, \theta_{S1}) = H\cos(90^\circ - |\theta_{S1}|) \quad (6)$$

$S0$ is the topmost servo, and $S1$ is the servo attached to $S0$. X_{off} and Y_{off} are the x and y components of the nozzle when $\theta_{S0} = -90^\circ$ and $\theta_{S1} = 0$. X_{NO} and Y_{NO} are the x and y components of the nozzle when $\theta_{S0} = 0$ and $\theta_{S1} = 0$. D and H are the y and z components, respectively, of the distance of the nozzle from when $\theta_{S1} = 0$ to when $\theta_{S1} = 90^\circ or - 90^\circ$, with $\theta_{S0} = 0$.

4 Methodology

The creation for the proposed 5-axis 3D printer can be divided into three major steps: 3-axis 3D printer assembly, development of g-code parser and transformer, and the addition of 4th and 5th axes on the printer using servos. The working 3-axis 3D printer is then modified into a 5-axis 3D printer through the use of servo pan/tilt system and some slight modification of its firmware. For this new printer's g-code to be compatible, an existing 3-axis g-code's x, y, and z values must then be extracted and transformed into a 5-axis system compatible g-code through the use of matrix transformations.

4.1 Hardware Materials

The 3D printer assembled is an open-source RepRap printer from RepRapPro, which is also an Ormerod 2 kit. This thesis can also be applied to other 3D printers provided the servo design is the same so that our software can be used and the firmware of the printer should support the M280 (set servo position) command. In his paper, Grutle used stepper motors for controlling the additional axes [5]. The researchers chose to go with servomotors because the 3D printer controller board that came with the printer does not support controlling additional stepper motors out of the box and requires purchasing an expansion board to do so.

The g-code generated can also be applicable to some printers because there is a standard set of g-code commands used by most printers. The g-code created by our parser only modifies existing g-code already generated by a slicer, so the slicer should generate g-code already applicable to the printer being used.

To test the 5-axis 3D printer, a PLA filament was used because of its ease of use. It is a relatively low-temperature material and does not exhaust toxic fumes when used in 3D printing. It does not require a very hot temperature for the heated bed. ABS filament was also used to create the additional parts of the printer for the added servos. These additional parts were placed near the hot end. ABS was used because it requires a higher temperature than PLA for it to melt.

To find the weight of the test prints, a weighing scale accurate to 1 g was used. The printers also reported the filament usage after every print which was recorded to know the amount of material used.

When the servos were mounted, a 600 mm-long linear shaft and a 600 mm aluminum profile were also used to increase the printable area of the printer.

4.2 *Software Programs*

Three different software programs were used for this thesis. Slic3r was used to convert 3D models to g-code instructions for the printer. A full-cloud CAD tool called OnShape was used to design and create new parts of the 3D printer to mount the sensors, the hot end, and the infrared sensor, as well as a new IR sensor bar. Lastly, Code::Blocks was used to develop the g-code parser and transformer.

4.3 *Development of G-Code Parser and Transformer*

Initially, the g-code transformer program was implemented using Python with the NumPy library, and after finding out that Python floating point operations seemed to be inaccurate, the program was rewritten to C++ with a matrix mathematics library called Eigen.

Nozzle Compensation Formula

Whenever the servomotors rotate, the nozzle gets placed in a different x, y, and z position. The x, y, and z distance relative to its original position can be expressed as a function of the two servos and the measured distances of the nozzle from its original position to various rotated positions. Subtracting the x, y, and z distances from the current x, y, and z position then allows the nozzle to compensate for its new position, moving the nozzle back to its original position while still maintaining its new rotation angles. The new nozzle position can be found using the following formulas based on Eqs. 4, 5, and 6:

$$X_{\text{previous}} = X_{\text{current}} - X_{NC} \tag{7}$$

$$Y_{\text{previous}} = Y_{\text{current}} - Y_{NC} \tag{8}$$

$$Z_{\text{previous}} = Z_{\text{current}} - Z_{NC} \tag{9}$$

4.4 Addition of 4th and 5th Axes on Printer

New parts were created using OnShape. The parts were printed in ABS using another 3D printer and installed onto the existing parts of the assembled 3D printer.

First, there is a servo mount on the X carriage (Fig. 2a) which holds the entire servo arm assembly. A second servo mount (Fig. 2b) attaches to the first servo. The second servo is attached to a servo arm which holds the hot end assembly and the IR sensor of the printer (Fig. 2c). Since the IR sensor is now located in a new position,

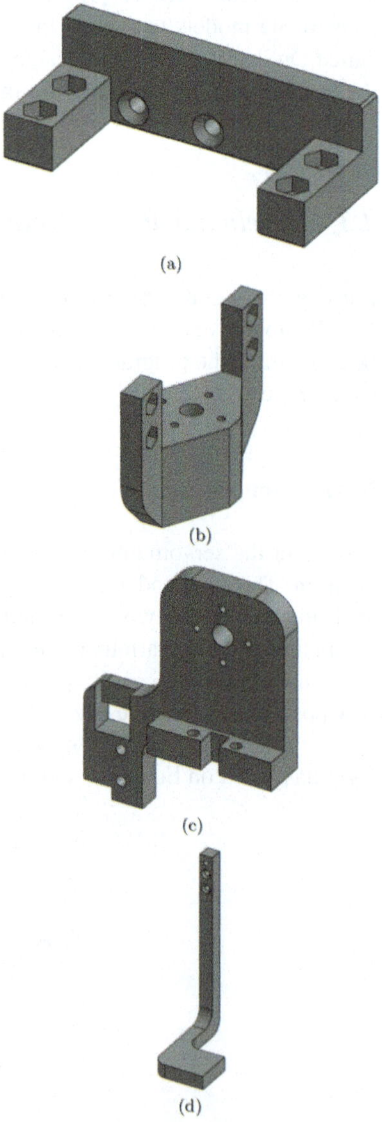

Fig. 2 3D-printed parts for the addition of the 4th and 5th axes (**a**) X carriage mount. (**b**) Servo arm. (**c**) Hot end and IR sensor mount. (**d**) IR sensor bar

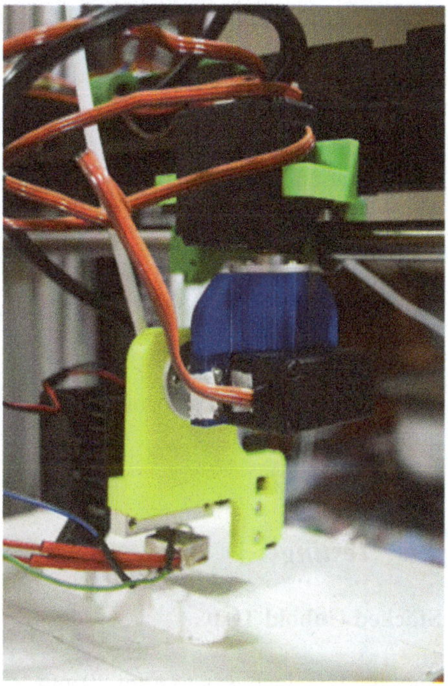

Fig. 3 Attached servos to 3D printer

the sensor bar (Fig. 2d) was redesigned so it attaches to one side of the x motor mounts. This is where the IR sensor detects its home position. The attached servos to the 3D printer can be seen in Fig. 3.

Firmware modifications were done using the Eclipse IDE for C/C++ Developers with the Arduino Eclipse plugin. It was modified to send the properly timed signal to the servos to a free pin we chose. The signal is sent whenever a M280 command is encountered. The servos also require a 5V signal and power supply to operate. A dual 5V and 3.3V breadboard power supply was purchased along with a 4-channel level shifter to shift 3.3V signals from the printer to the servos.

The Ormerod printer has a configuration file present in its SD card which can be modified rather than having to recompile the entire firmware when changing settings. For the printer, the homing procedure had to be changed to rotate the servos in certain positions before homing. The maximum speed of the Z axis was increased as well.

4.5 Running the 5-Axis 3D Printer

Firstly, 3D models with the STL file format are sliced using Slic3r. Each part that has a different rotation than normal operation must be sliced as separate files and exported to separate g-code files. The g-codes generated have two profiles: a standard printing profile and a profile for the g-code transformer. The latter simply has the initial "skirt" removed, and the center of the bed is set as the bottom center of the model.

Afterward, the g-code transformer is run. There are some user-changeable settings in the program with defaults for the printer. This is where the user can input what angles the servos are set to or its position translation in relation to the center of the bed. The program asks for the input file to read and output file to write. The input file should be the g-code with the nonstandard print profile. The output g-code file contains the necessary rotations and translations for the original g-code's coordinates as well as the M280 g-code commands for setting the printer's servo rotation.

After transforming all the necessary g-code, the user must upload the separate g-code files to the printer and run the files in sequence until the whole object is printed.

4.6 *Testing*

Stacked Cuboid Test

The first test done was to print 20 mm × 20 mm × 10 mm cuboid with $\theta_{S0} = 90°$ and $\theta_{S1} = 0$. For the next section, θ_{S0} is set to $-90°$ (effectively rotating the nozzle along the Z axis), and then another cuboid of the same size gets printed on top of the first one (Fig. 4a). This test should demonstrate the precision of the output transformed g-code file.

Attach Overhanging Cube on the Side of a Tall Tower

The second test involved printing a tall 20 mm × 20 mm × 100 mm tower with $\theta_{S0} = 90°$ and $\theta_{S1} = 0$. When the tall tower is done printing, a small 20 mm × 20 mm × 10 mm cuboid is printed on the side of the tower with the nozzle rotated 90° (Fig. 4b).

45° Overhang Test

A tall 20 mm × 20 mm × 100 mm tower is printed again, and a 20 mm-tall solid with a 20 mm × 20 mm base with a 40 mm × 40 mm top area (creating a 45° overhang) gets printed on the side of the tower with the nozzle rotated 90°. The top area of the solid is extruded outward an additional 20 mm (Fig. 4c).

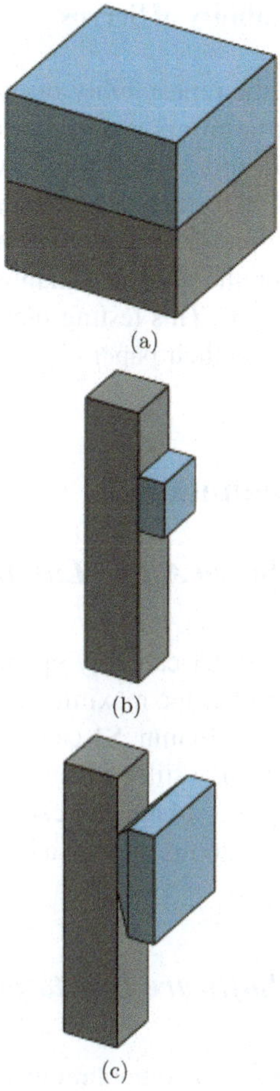

Fig. 4 CAD models of all test prints. (**a**) Stacked cuboid test. (**b**) Tall tower with overhanging cuboid test. (**c**) 45° overhang test

Print Time

The time it takes to print without supports using the 5-axis 3D printer and with supports using the default setting of the 5-axis 3D printer, where the additional 4th and 5th axes aren't used, is measured. Other printers can be used to print the parts with support as a basis of comparison.

Material Usage

The amount of material (measured using weight) used to print without supports using the 5-axis 3D printer and with supports using the default setting of the 5-axis 3D printer, where the additional 4th and 5th axes aren't used, are measured.

Repeatability of Servos

To test the repeatability of the servos, first the angle of the servo is set to zero and then an arbitrary angle is set and the angle is set back to zero. This is repeated multiple times. And the servo is checked if it goes back precisely to the angle specified. To check, the distance from the initial point to the final point is measured. Doing this for several trials and taking the maximum distance give the maximum error for the servo mechanism. Testing is done to ensure reliability of the servo mechanism. This testing method was based on the work of Cando, Dy-Reyes, and Mariñas in their paper [7].

5 Limitations

5.1 Servo Angle Limitations

At $Z = 0$, S1 can only operate from $-90°$ to 0, depending on θ_{S0}. The higher θ_{S0} is, the higher the maximum angle is of the operational range of S1.

At $Z \geq 14$ mm, S1 can operate from -90 to $90°$ without colliding with the bed, but there may still be instances where S1 can collide with the x arm.

At $Z = -14$ mm, S1 can operate at $-90°$ without colliding with the bed provided that a rotation is done at a higher z position first before moving to $Z = -14$ mm.

5.2 Software Limitations

For the speed computations in the g-code files being generated by Slic3r, the speed per layer is dependent on the estimated time per layer. For the proposed printer, the z movement is slow, and it takes time per layer. This causes the x and y movement to go at full speed causing the nozzle to extrude sparsely.

Slowing down the print speed will cause the entire layer to be finished in more time, and printing the entire 3D object will take longer than just printing with support. Increasing Z speed is a better solution for our printer because it can lessen the printing time, but it causes more violent wobbling and makes the whole z movement unstable. Applying a speed parameter to every non-z movement with the extrusion in the g-code file is a possible fix. This allows the x and y speed to go slower, while the z movement is set to the fastest. This works better if z speed is also increases.

5.3 Limitations in Simultaneous Rotation and Movement

The rotation is done separately from the movement. It is possible to do both but would need a major rework of the g-code transformer.

5.4 Manual Process for Calibration

The researchers did not have access to professional tools that can be used for calibrating the machine. Everything had to be done manually.

6 Results and Discussions

6.1 Testing Results

Stacked Cuboid Test

The resulting print of the stacked cuboid test was slightly misaligned by 0.65 mm as seen in Fig. 5a. This can be resolved by changing the nozzle offset setting in the g-code transformer.

Attach Overhanging Cube on the Side of a Tall Tower

During testing, the printing of the small cuboid on the side failed due to misalignment. This was resolved by restarting the print and holding the tall tower in place. The print still successfully finished as seen in Fig. 5b.

45° Overhang Test

Printing of this piece took a long time to print for 3 h and 48 min; however printing was successful as seen in Fig. 6a (left).

Print Time

Using the 45° overhang test, the time it takes to print it with supports using 3 axes versus printing it without supports using 5 axes was measured and compared. It can be seen from Table 1 that the Creator Pro tales longer to print with supports than without supports. However with the Ormerod, it takes longer to print without

Fig. 5 Test prints. (**a**) Stacked cuboid test. (**b**) Tall tower with overhanging cuboid test

(a)

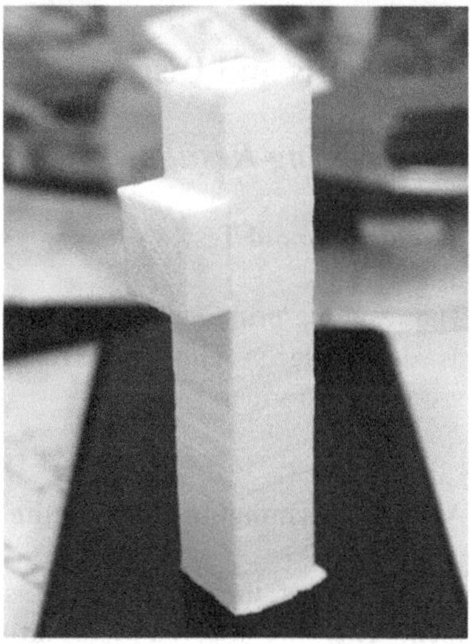

(b)

supports than with supports. This is due to the slow z movement that slows down the entire print and increases print time substantially.

Material Usage

The 45° overhang test was used again. Each print was measured using a weighing scale along with the filament usage reports. Both printers consume less filament when printing without supports than with supports. From Table 1, for the Ormerod printer, the support structures use 7 grams of material or 2.345 meters of filament; the Creator Pro uses 8 grams of material or 2.81 meters of filament.

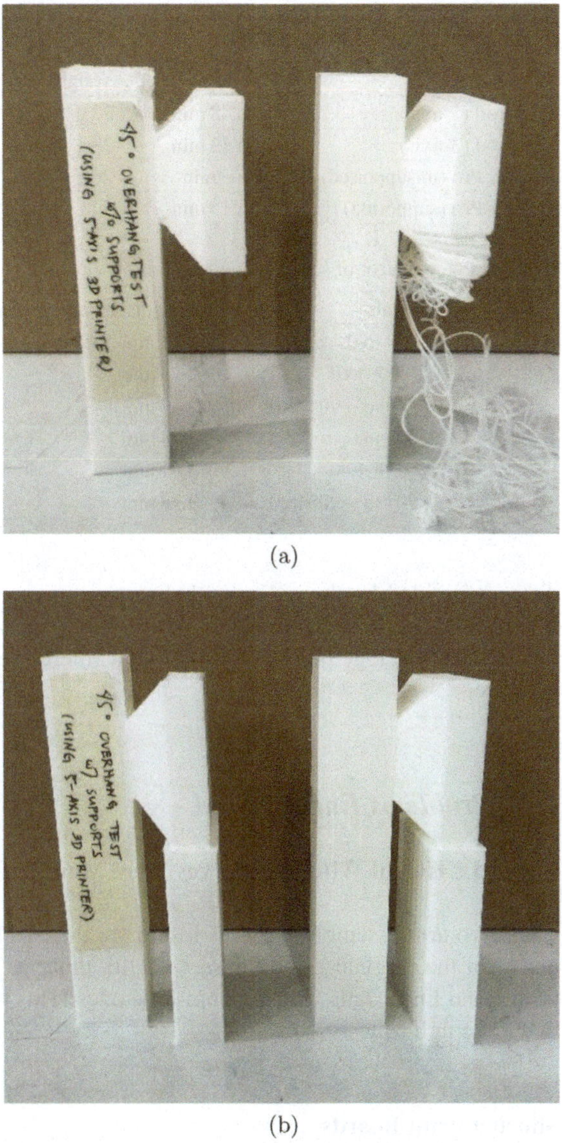

Fig. 6 Side view of test prints from the Ormerod and Creator Pro. (**a**) Unsupported prints from the Ormerod (left) and from the Creator Pro (right). (**b**) Supported prints from the Ormerod (left) and from the Creator Pro (right)

Repeatability of Servos

A 150 mm servo arm was printed and mounted on the servos as a type of servo horn. For every trial seen on Table 2, each servo was rotated to 90° then back to 0 five times. After every rotation, a point was marked at the very tip of the servo arm. After doing the set of trials, maximum distance between points was measured. This distance was the variance of the rotation with a radius of 151.42 mm. The measured

Table 1 Table of printing times, weight of print, filament usage, and material cost of the 45° overhang test print on various printers

	Time	Weight	Filament used	Cost
Ormerod (5-axis)	3 h 48 min	22 g	7461.1 mm	₱24.87
Ormerod (3-axis)	2 h 13 min	29 g	9806.1 mm	₱32.69
Creator Pro (unsupported)	2 h 3 min	26 g	9071.5 mm	₱30.24
Creator Pro (supported)	2 h 13 min	34 g	11,878 mm	₱39.59

Table 2 Percent error of servo rotation in 15 trials

	Trials	Max distance between points	Percent error
S1	90° to 0	2.65 mm	0.27853%
	−90° to 0	2.75 mm	0.28904%
	90° to −90° to 0	2.37 mm	0.24911%
S2	90° to 0	1.24 mm	0.13033%
	−90° to 0	2.61 mm	0.27433%
	90° to −90° to 0	2.78 mm	0.29220%

distance divided by the radius is the percent error when rotating a certain radius. The average percent error for the three trials for S1 is 0.27%, while for S2 it is 0.23%. Both average percent errors are low; therefore the servos can be considered accurate.

6.2 *Problems Encountered*

Limited *z* Height When the Servos Were Mounted

The servo arm assembly made us lose more than 100 mm of the available printing height in the original printer design. To fix this, we had to replace the aluminum profile and linear rails with 600 mm-long ones. This problem was also encountered by Grutle in his thesis [5].

Shorted Print Boards

The printer controller boards caused a short when the nozzle crashed into the bed and made contact with the aluminum section of the bed. Two boards were fried because of this. To fix this problem, we wrapped the aluminum sections of the bed with Kapton tape, electrically insulating it.

7 Conclusion

A working 3-axis 3D printer was modified in to a 5-axis 3D printer through the use of servo motion system and modification of its firmware. G-code files from a g-code transformer which was developed by the researchers contain the g-code commands that have been made compatible with the new 5-axis 3D printer system. The 5-axis 3D printer successfully saves filament by eliminating the need for support structures when printing certain geometry. However, the system takes a long time to do so.

For future study in this thesis, a more rigid 3D printer with a better linear motion system is recommended for use with a 5-axis system due to its greater tolerance with vibrations or small movements should be better. More sophisticated calibration methods are also recommended.

Acknowledgements The authors would like to acknowledge the contributions and help of their thesis advisers, Carlos M. Oppus and Juan Antonio G. Mariñas, and their professors in the Ateneo de Manila University, Dr. Luisito Agustin, Daniel M. Lagazo, and Dr. Rosula S.J. Reyes, for the input they provided for the betterment of this thesis.

References

1. Gero, J., & Kelly, N. (2015). How to make CAD tools more useful to designers through re-representation. Key Centre of Design Computing and Cognition. University of Sydney, Australia.
2. Pirjan, A., & Petroanu, D. (2013). The impact of 3D printing technology on the society and economy. *Journal of Information Systems & Operations Management, 7*(2), 360–370. Romanian-American University.
3. About – RepRap. http://reprap.org/wiki/About. Last Accessed 21 Oct 2015.
4. Makhanov, S., & Anotaipaiboon, W. (2007). *Advanced numerical methods to optimize cutting operations of five-axis milling machines* (pp. 35–38). Berlin/Heidelberg: Springer.
5. Grutle, Ø.K. (2015). *5-axis 3D printer*. University of Oslo.
6. Dunn, F., & Parberry, I. (2012). *3D math primer for graphics and game development*. Boca Raton, FL: CRC Press.
7. Cando, J., et al. (2013). *3D CNC milling machine built from locally available materials*. Ateneo de Manila University.

Part IV
Internet of Things, ICT and Artificial Intelligence

Alphanumeric Test Paper Checker Through Intelligent Character Recognition Using OpenCV and Support Vector Machine

Jessica S. Velasco, Anthony Aldrin V. Beltran, Joie Ann C. Alayon, Paul Edgar B. Maranan, Cheza Marie A. Mascardo, Justine Mae B. Sombrito, and Lean Karlo S. Tolentino

1 Introduction

Nowadays, there is a test paper checker machine commercially available in the market, but it is only used for large numbers of examinees, like in nationwide examinations. Those machines are only capable of checking multiple choice questions because it enables standardization, and these types of exams can be graded quickly. They use optical mark recognition (OMR), a process used to get information through a marked field and have a specially designed answer sheet [1]. However, acquiring and operating these machines can be quite expensive, so past studies offer another method using digital image processing techniques for hand marked answers and it will be applied in a local computing machine [2–6].

A device was developed that is capable of reading and checking the answers written in the alphanumeric character form and will give a reliable result of the scores received by the examinees. It will also have an item analysis that will determine the difficulties of each item on the given examinations. With this device, the examiner and the examinee will also be given an immediate score after the examination as the device can store data and results in the database included in the system and it can produce a printout of the results that will be given to the

J. S. Velasco (✉) · L. K. S. Tolentino
Department of Electronics Engineering, Technological University of the Philippines, Manila, Philippines

Center for Engineering Design, Fabrication, and Innovation, College of Engineering, Technological University of the Philippines, Manila, Philippines
e-mail: jessica_velasco@tup.edu.ph

A. A. V. Beltran · J. A. C. Alayon · P. E. B. Maranan · C. M. A. Mascardo · J. M. B. Sombrito
Department of Electronics Engineering, Technological University of the Philippines, Manila, Philippines

A. Beltran Jr. et al. (eds.), *World Congress on Engineering and Technology; Innovation and its Sustainability 2018*, EAI/Springer Innovations in Communication and Computing, https://doi.org/10.1007/978-3-030-20904-9_9

examinees. The machine can be used for two types of questions: multiple choice and problem solving. In a multiple-choice question, letters A–E are only applicable to use, while in problem solving, the machine is only capable of reading numeric characters and asterisk that corresponds to a decimal point. The answers must be written on the specific fields for the machine to recognize it accurately.

2 Machine Vision and Learning Techniques

The first step in handwriting recognition systems is the preprocessing, which helps to reduce the inconsistency of handwriting and enhance the accuracy for segmentation and recognition methods. Feature extraction is the second step to the systems of handwriting recognition and classification is the last step. Histogram of Oriented Gradient (HOG) descriptor is a type of feature extraction, which is used for improved accuracy; it is also a successful descriptor for detection [7].

Support Vector Machines (SVM) in various modern learning applications are applied successfully by many researchers such as Optical Character Recognition (OCR), bioinformatics, document analysis, and image classification. SVM is the most important and most effective classifier there are set of supervised learning methods used for classification, regression and outlier detection [8–12].

Incorporating human capability in computers has been receiving attention in the research field. And over the years, humans have been thinking of developing an algorithm that can read and interpret printed textual documents, so that they can be automatically converted into an alternate medium or format. Such a technology has been developed. The system is named as Optical Character Recognition (OCR) [13].

3 Methodology

3.1 Hardware Development

Figure 1 shows the 3D and actual representation of the proposed device. It has a dimension of 16.25″ in width, 13.78″ in height, and 12.91″ in length. The PC tablet is for the user interface, database, and results of the processed data. The scanner is used for scanning of data to be processed by the intelligent character recognition. The thermal printer is for the printout of the examination results and the sintra board is used for the casing of the device. Also, a USB hub is included for the user to have easy access to the database on the system.

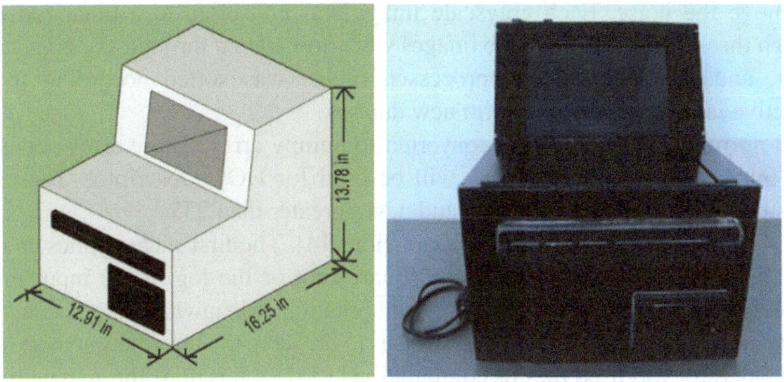

Fig. 1 3D design of the device

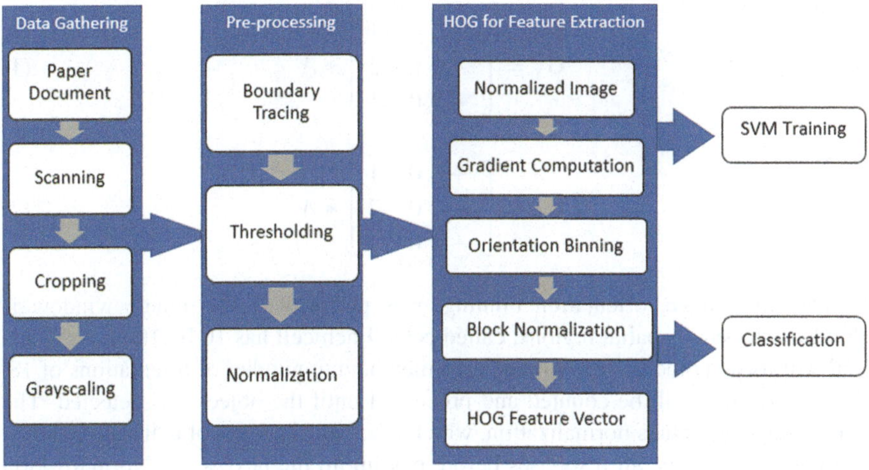

Fig. 2 Program flow of SVM and HOG feature extraction

3.2 Software Development

In this study, SVM was used for classification. Histogram of Oriented Gradients (HOG) was used along with SVM for its algorithm and feature extraction. Figure 2 shows the program flowchart used in this study. HOG is a feature descriptor found in the Scikit-Image and OpenCV. HOG is used to detect character in an image through counting occurrences of gradient orientation in localized portions of image and computing the dense grid of uniformly spaced cells and uses overlapping local contrast normalization. Python was used as programming language, which is commonly used for image processing and computer vision. The images were cropped to extract each image of a character through edge detection and boundary tracing and converted to a grayscale image. These images underwent preprocessing to remove the noise. Each grayscale image was converted to a binarized image through thresholding. And all the images were normalized through skewing, center-extent, and resizing. Each preprocessed image was sorted according to their respective labels and saved to form new dataset.

All normalized images were converted to numpy array. These arrays consist of pixel intensities of images, which will be used for HOG Descriptor. These pixel intensities will never be less than 0 and never greater than 255.

HOG was performed for feature extraction [14]. The first stage applies an image normalization through computing the square root or the log of the input images. It is designed to lessen the effects of illumination. Meanwhile, the second stage computes first order image gradients. This is to preserve the information in an image. Sobel Operator was used for this computation [15–18]. It is done through getting the horizontal derivative (1) and the vertical derivative (2) of the image using 3x3 Sobel masking and using a 2-D digital signal processing convolution operation.

$$G_x = \begin{bmatrix} +1 & 0 & -1 \\ +2 & 0 & -2 \\ +1 & 0 & -1 \end{bmatrix} * A \tag{1}$$

$$G_x = \begin{bmatrix} +1 & 0 & -1 \\ +2 & 0 & -2 \\ +1 & 0 & -1 \end{bmatrix} * A \tag{2}$$

The third stage, orientation binning, was performed. The image window is divided into small spatial regions, called cells. Each cell has 10 by 10 pixels. Each cell will form a block. This will be performed having number of orientations of 18. The bin formed will be counted and processed until the object was detected. The fourth stage computes normalization, which takes local groups of cells and contrast normalizes their overall responses before passing to the next stage. Normalization makes the images processed to be prepared for the training. L2-Hys was used for block normalization, which is derived from L2-Norm (3), but limiting the maximum values of unnormalized descriptor vector, V, and renormalizing it [15]. Variable "e" represents small-value constants. Finally, the resulting HOG feature vector is produced and can be used for both SVM training and classification.

$$f = \frac{V}{\sqrt{||V_2^2|| + e^2}} \tag{3}$$

Figure 3 shows the flowchart of the proposed Test Paper Checker Machine. The scanned image is cropped and extracted by the Histogram of Oriented Gradient (HOG) algorithm. The Support Vector Machine (SVM) is used as Machine learning for recognition. Python is used as the programming language of the HOG and SVM. The process of checking, database management, and printing the result is done using Visual Basic.

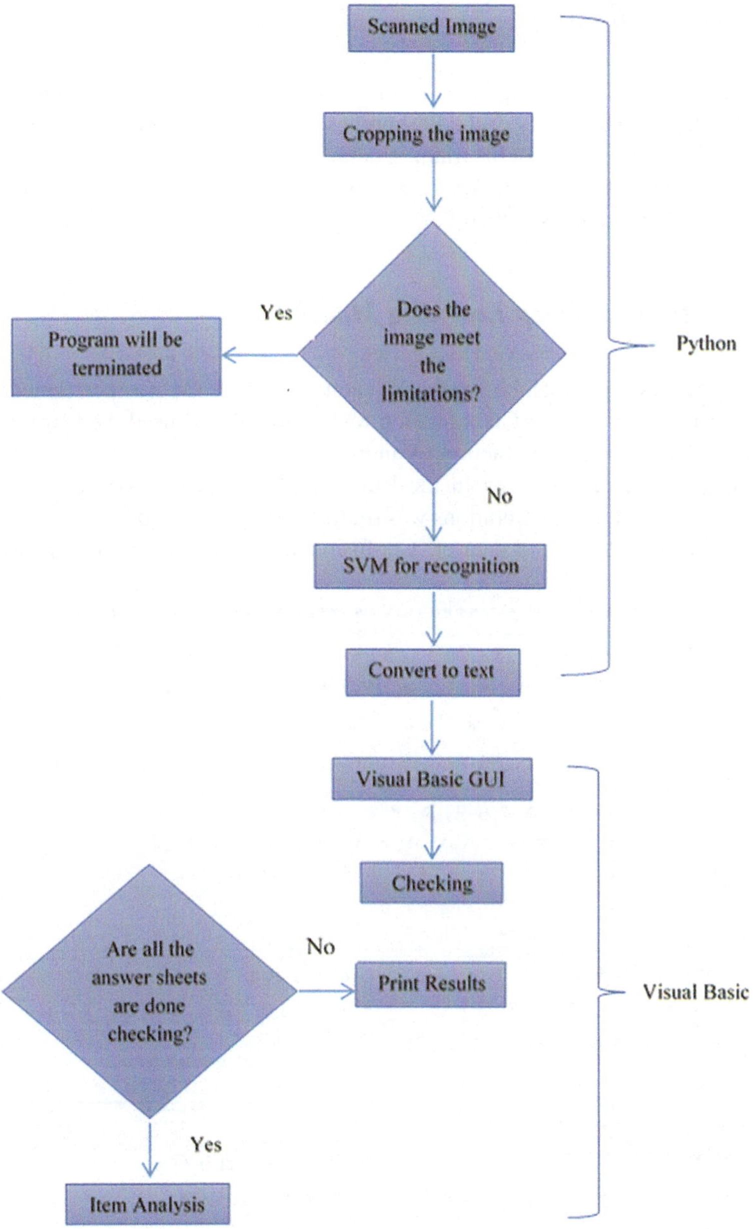

Fig. 3 Program flow of SVM and HOG feature extraction

3.3 Data Gathering

Handwritten samples were gathered from the students of the College of Engineering in Technological University of the Philippines – Manila, which produced 500–1000 data sets that can be used for the training data. The trained data are used for machine learning to have a high accuracy in handwritten character recognition. Figure 4 shows one of the samples gathered.

3.4 Training of Deep Learning Model

After HOG Descriptor, which produced feature vector, SVM was performed. SVM used linear Support Vector Classification for training. SVM used the feature vector produced from feature extraction. Random State was set to 42 for improved accuracy. Parameters were set in the Linear SVC found in scikit-learn that will perform SVM. Number of iterations was set to 1000. The proponents set the number of test data to 20% of the dataset. Confusion matrix of the trained dataset was

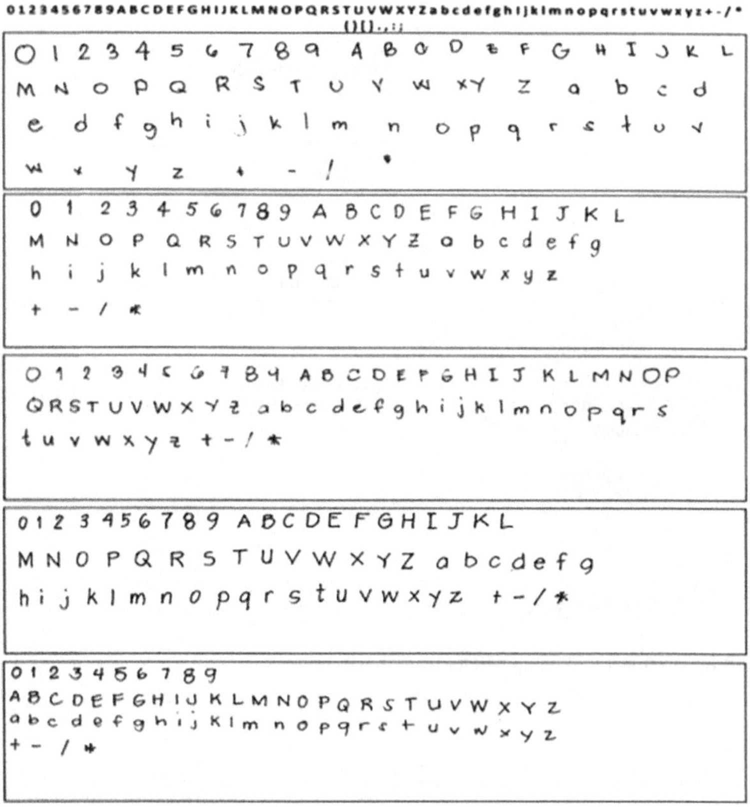

Fig. 4 Handwritten characters for data set

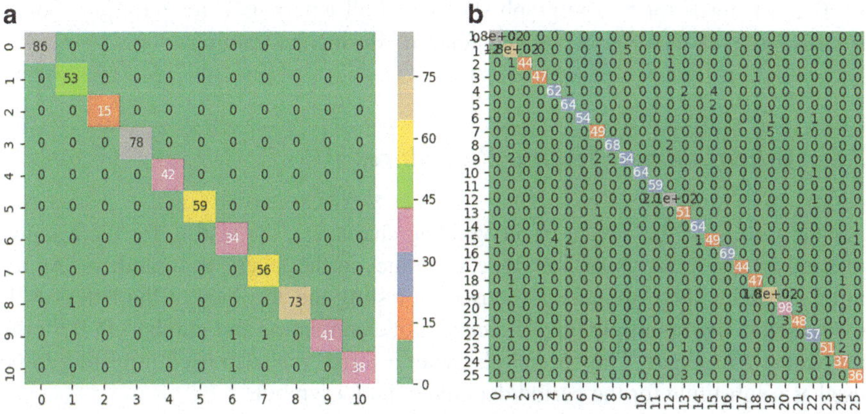

Fig. 5 (**a**) Confusion matrix of numbers and decimal point. (**b**) Confusion matrix of alphabet characters

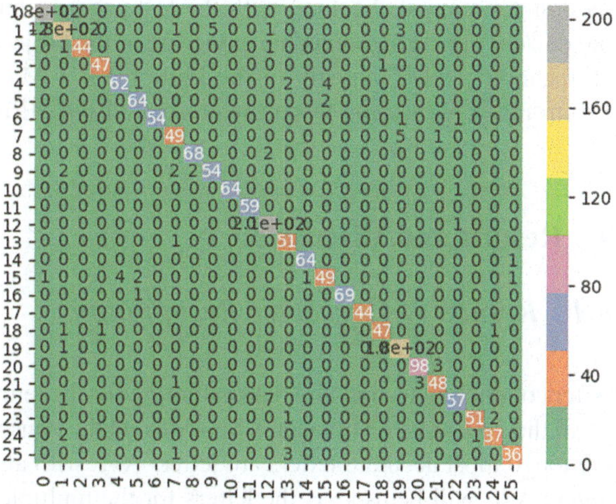

Fig. 6 Confusion matrix of letters A–E

produced and plotted with Matplotlib project of Python. The y-axis represents the true label and the x-axis represents the predicted label. The diagonal boxes represent the number of correct predictions of the label. The numbers outside the diagonal boxes were the number of incorrect predictions of the label.

Figure 5a shows that all numbers were recognized correctly except 4 images during the training. Number 8 was recognized as number 1, Number 9 as number 6 and number 7, and the decimal point was predicted as number 6. Meanwhile, Fig. 5b shows the confusion matrix of alphabet characters. It shows that there are several letters that were predicted incorrectly during the training. All the letters in the x- and y-axis represent each alphabet such as letter A for 0 and letter z for 25. The letters with the most number of incorrect predictions were W and H. Letter A had the most number of correct predictions. When the letters used are reduced, it resulted to a higher accuracy in recognizing characters as shown in Fig. 6. All the numbers in the x- and y-axis represent each letter used for MCQ, such as letter A for 0 and letter E for 5. The graph shows that all letters used for multiple choice, such as letters A–E were recognized correctly during the training.

3.5 Validation and Testing Procedures

For the testing of the device in recognizing characters, the testing for the training of handwriting recognition was done 100 times in the letters and numbers. As for more accuracy, the respondents focused on testing the accuracy of the letters A–E to be used in their multiple type choice part of examination and numbers 0–9 for the problem-solving part. The special character asterisk was also used in testing to be used as a substitute for the period/decimal point (.) symbol.

As for the data gathering, 130 students were gathered to test their device. Examinations were with a total of 100 questions to be answered. Two ways of checking were performed for comparing of score as well as data. The first way was by doing a manual checking with the test papers, and the other way was by using the device. The testing and data gathering of the project study device was conducted at Technological University of the Philippines – Manila, College of Engineering, Electronics Engineering Department, Ayala Blvd., Ermita, Manila, Philippines.

4 Results and Discussion

4.1 Results of Character Recognition

After conducting testing and data gathering for the letters and numbers, the accuracy of all the 26 English Alphabet is 86.45% and the consecutive 5 letters that belong to the top10 most accurate letters are letters A–E with an accuracy rate of 95.4%, thus letters A–E will be used as the letters for the multiple-choice type of examination. As for the accuracy of the numbers, it has a 91.6% of accuracy. Using asterisk as a substitute for the period symbol was also decided. Table 1 shows the percentage accuracy for both alphabets and numeric characters.

From the 130 respondents, 121 of them are in the range in relation to the manual checking and do not recognize any error in recognizing characters. Table 2 shows how many times the device correctly and incorrectly recognized the characters written. In the result, this analysis came up with 93.0769% of accuracy. Applying SVM as ICR has improved and increased the accuracy of the device. Neat and

Table 1 Percentage accuracy for alpabets and numbers recognition

	Percentage accuracy (%)
Alphabets	86.45
Numbers	91.6

Table 2 Correctly and incorrectly recognized characters

	Number of times the device recognized the characters written
Correct	121
Incorrect	9

legible handwriting can be accurately processed by the device. With the method used, the device can check within 57.1111 s per answer sheet with 120 items.

5 Conclusion

A test paper checker was developed in this study that will recognize a handwritten text using Intelligent Character Recognition (ICR) for Alphanumeric Characters, which is based on OpenCV for image processing and SVM for classification. It has a 93.0769% accuracy of correct recognition of characters. For future work, deep learning algorithms such as convolutional neural networks will be implemented.

References

1. Catalan, J. A. (2017). A framework for automated multiple-choice exam scoring with digital image and assorted processing using readily available software. In *DLSU research congress 2017* (pp. 1–5). Manila: De La Salle University.
2. China, R. T., de Assis Zampirolli, F., de Oliveira Neves, R. P., & Quilici-Gonzalez, J. A. (2016). An application for automatic multiple-choice test grading on android. *Revista Brasileira de Iniciação Científica, 3*(2), 4–25.
3. Muangprathub, J., Shichim, O., Jaroensuk, Y., & Kajornkasirat, S. (2018). Automatic grading of scanned multiple choice answer sheets. *International Journal of Engineering and Technology(UAE), 7*(2), 175–179.
4. Čupić, M., Brkić, K., Hrkać, T., Mihajlović, Z., & Kalafatić, Z. (2014). Automatic recognition of handwritten corrections for multiple-choice exam answer sheets. In *2014 37th international convention on information and communication technology, electronics and microelectronics, MIPRO 2014 – proceedings* (pp. 1136–1141). Opatija: IEEE.
5. Fisteus, J. A., Pardo, A., & García, N. F. (2013). Grading multiple choice exams with low-cost and portable computer-vision techniques. *Journal of Science Education and Technology, 22*(4), 560–571.
6. Abdul Nabi, A. H., & Aljarrah, I. A. (2016). An automated multiple choice grader for paper-based exams. In P. Soh, W. Woo, H. Sulaiman, M. Othman, & M. Saat (Eds.), *Advances in machine learning and signal processing. Lecture notes in electrical engineering* (Vol. 387). Cham: Springer.
7. Kadhm, M. S., & Hassan, A. K. A. (2015). Handwriting word recognition based on SVM classifier. *International Journal of Advanced Computer Science & Applications, 6*(11), 64–68.
8. Fu, W., & Menzies, T. (2017). Easy over hard: A case study on deep learning. In *Proceedings of the 2017 11th joint meeting on foundations of software engineering* (pp. 49–60). Paderborn: ACM.

9. Hamid, N.A., Sjarif, N.N.A. (2017) Handwritten recognition using SVM, KNN and Neural Network. arXiv preprint arXiv:1702.00723, 1–11.
10. Kaensar, C. (2013). A comparative study on handwriting digit recognition classifier using neural network, support vector machine and K-nearest neighbor. In P. Meesad, H. Unger, & S. Boonkrong (Eds.), *The 9th international conference on computing and information technology (IC2IT2013). Advances in intelligent systems and computing* (Vol. 209). Berlin, Heidelberg: Springer.
11. Tolentino, L. K. S., Orillo, J. W. F., Aguacito, P. D., Colango, E. J. M., Malit, J. R. H., Marcelino, J. T. G., Nadora, A. C., & Odeza, A. J. D. (2017). Fish freshness determination through support vector machine. *Journal of Telecommunication, Electronic and Computer Engineering, 9*(2-5), 139–143.
12. David, L. C. G., & Ballado, A. H. (2015). Mapping mangrove forest from LiDAR data using object-based image analysis and Support Vector Machine: The case of Calatagan, Batangas. In *2015 international conference on humanoid, nanotechnology, information technology, communication and control, environment and management (HNICEM)* (pp. 1–4). Cebu City: IEEE.
13. Meher, S., & Basa, D. (2011). An intelligent scanner with handwritten odia character recognition capability. In *2011 fifth international conference on sensing technology* (pp. 53–59). Palmerston North: IEEE.
14. Dalal, N., & Triggs, B. (2005). Histograms of oriented gradients for human detection. In *2005 IEEE computer society conference on computer vision and pattern recognition (CVPR'05)* (pp. 886–893). San Diego, CA: IEEE.
15. Sobel, I. (2014). *History and definition of the so-called "Sobel operator", more appropriately named the Sobel-Feldman operator* (pp. 3–4).
16. Han, D. S., Serfa Juan, R. O., Jung, M. W., Cha, H. W., & Kim, H. S. (2017). Development of a novel fast rotation angle detection algorithm using a quasi-rotation invariant feature based on Sobel edge. *Journal of Telecommunication, Electronic and Computer Engineering, 9*(2-6), 33–36.
17. Tolentino, L. K. S., & Beleno, D. M. T. (2017). Development of a 3D disparity estimation processing architecture. *International Journal of Applied Engineering Research, 12*(19), 8420–8424.
18. Ibarra, J. B., Paglinawan, A., Sejera, M., Dema-ala, F. A., Enriquez, M., Glodo, J. R., & Marty, F. P. (2017). Measurement of overall decentration, angle deviation, and prism diopters in categorized strabismus cases using mathematical morphology algorithm. In *2017 IEEE 9th international conference on humanoid, nanotechnology, information technology, communication and control, environment and management (HNICEM)* (pp. 1–4). Manila: IEEE.

Automated Water Quality Monitoring and Control for Milkfish Pond

Shiella Marie P. Garcia, Cyd Laurence B. Santos, Karen Mae E. Briones, Sean Michael L. Reyes, Maurice Alyana G. Macasaet, and Rolando Pula

1 Introduction

Milkfish is one of the largest volumes of fish traded each day in almost all public markets in the Philippines. According to the Food and Nutrition Research Institute, about one-fifth of the 25 kg of fresh fish consumed by Filipinos each year is "Bangus" or 4 kg of milkfish per person per year [1]. Milkfish is an important commodity that is widely cultured in the Philippines. It remains to be an essential or staple food, with production growing at a rate of 8.5% per annum. Milkfish is in high demand and widely distributed in the Indo-Pacific region and is increasing yearly as the consumer population increases [2].

S. M. P. Garcia (✉)
Computer Engineering Department, University of Perpetual Help System Dalta, Las Piñas Campus, Las Pinas, Metro Manila, Philippines
e-mail: smpgarcia1972@gmail.com

C. L. B. Santos · K. M. E. Briones · S. M. L. Reyes · M. A. G. Macasaet
Computer Engineering Department, University of Perpetual Help System Dalta, Las Piñas Campus, Las Pinas, Metro Manila, Philippines

College of Engineering, University of Perpetual Help System Dalta, Las Piñas Campus, Las Pinas, Metro Manila, Philippines
e-mail: cydlaurencesantos@gmail.com; karenmaebriones@gmail.com; sean_reyes332@yahoo.com; magm014@yahoo.com

R. Pula
Graduate Studies, Mapua University, Manila, Philippines

School of Graduate Studies, Mapua University, Manila, Philippines
e-mail: pularolando@gmail.com

A. Beltran Jr. et al. (eds.), *World Congress on Engineering and Technology; Innovation and its Sustainability 2018*, EAI/Springer Innovations in Communication and Computing, https://doi.org/10.1007/978-3-030-20904-9_10

The Philippines as of today imports more than 50% of its 2 billion fry annual requirement from Indonesia because of the lack of innovativeness in terms of implementing technology to help the Milkfish industry [3].

According to Freshwater Aquaculture, when water quality depreciates, milkfish survival is affected, and ultimately massive fish kills may occur. Water quality is a critical factor when culturing any aquatic organism. The quality of the water in the production systems can significantly affect the organism's health and the costs associated with getting a product to the market. Optimal water quality varies by species and must be monitored to ensure growth and survival.

To keep the water properties within safe levels, one must understand those processes so that the elements inhibiting growth and survival can be detected and their impact can be minimized [4]. The water quality parameters, such as ammonia, dissolved oxygen, pH level, temperature, and turbidity are commonly monitored in the aquaculture industry. Ammonia is the most important water quality parameter next to oxygen which affects fish. Exposure to low levels of ammonia causes bacterial infections for fishes [5]. Together with ammonia, low levels of dissolved oxygen are most frequently associated with hot, cloudy weather, algae die-offs [6]. On the other hand, higher temperature increases oxygen demand for fishes in the water and causes decreased solubility of oxygen and also increased level of ammonia in water [7]. Other parameters that greatly affects water quality are pH and turbidity of water. Extreme pH levels or when pH changes rapidly also affects animals to become stressed and die which is why control of these parameters are necessary to maintain the water quality [8]. In addition to this, excessive suspended sediment can impair water quality for aquatic and human life, impede navigation, and increase flooding risks which can be a result of turbidity in water not being controlled or maintained [9]. Each water quality parameter alone can directly affect the animal's health. Exposure of fish to improper levels of dissolved oxygen and ammonia leads to stress and disease. However, in the complex and dynamic environment of aquaculture ponds, water quality parameters also influence each other. Unbalanced levels of temperature and pH can increase the toxicity of ammonia. Thus, maintaining balanced levels of water quality parameters is fundamental for both the health and growth of culture organisms. It is recommended to monitor and assess water quality parameters on a routine basis [10].

2 Methodology

2.1 Conceptual Paradigm

Input Process Output (IPO) was followed in the study. Input parameters are the pH, Temperature, DO, Turbidity, and Ammonia. Measured values for pH, Temperature, DO, Turbidity, and Ammonia of the water will be compared based on the optimum values. These values will be processed/compared on the optimum values of the parameters. The system will then identify if the values of the measured parameters

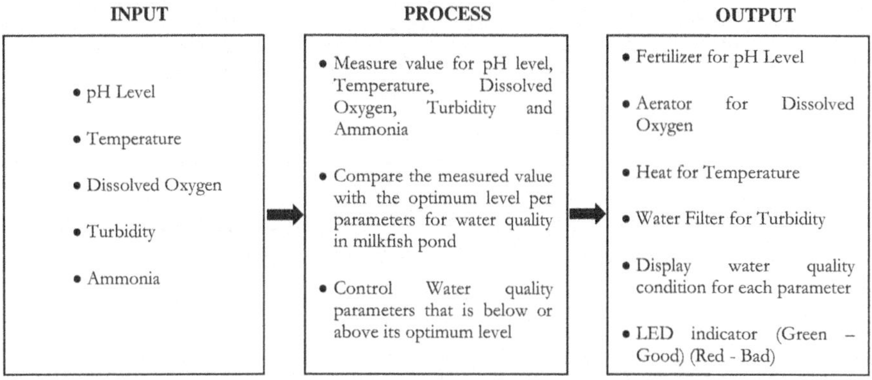

Fig. 1 Conceptual framework

Table 1 System condition for optimum level of measured parameters

Parameter	Optimum level	Conditions	LED color display
Ammonia	≤0.02 ppm	As long as the temperature and pH are at optimum level, the condition of ammonia is still good	≤0.02 ppm – Green LED >0.02 ppm – Red LED
Dissolved oxygen	≥3 ppm	If dissolved oxygen level is less than 3 ppm (<3 ppm), aerator will turn on	≥3 ppm – Green LED <3 ppm – Red LED
pH	6.8–8.7	If pH level is less than 6.8 (<6.8), acid fertilizer will release; if it is greater than 8.7 (>8.7), it will release alkaline fertilizer	<6.8 – red LED = 6.8–8.7 – green LED >8.7 – Red LED
Temperature	22 °C–35 °C	If temperature is less than 22 °C (<22 °C), the heater will turn on; if it is greater than 35 °C (>35 °C), aerator will turn on	<22 °C – Red LED = 22 to 35 °C – Green LED >35 °C – Red LED
Turbidity	3.8 V	If turbidity is less than 3.8 V (<3.8 V), filter will turn on	≥3.8 V – Green LED <3.8 V – Red LED

are below or above the optimum values. The system automatically controls the values of the measured parameters (pH, DO, Turbidity, and Ammonia), [7] if the measured values are below or above the range of the optimum values. Figure 1 shows the Conceptual Framework of the study, while in Table 1, you can find the optimum level/range of each measured parameter for normal values.

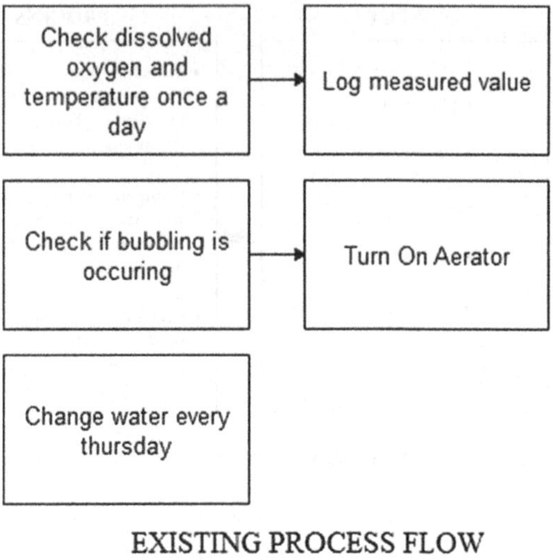

Fig. 2 Existing operation based on the interview of the fishpond owners

2.2 Description of the Existing System from the Study Area

Based from an interview conducted by the proponents in Rock Fin Fish Farm, a farmer is assigned to monitor the water quality of their ponds by manually checking the dissolved oxygen and temperature. This is done once a day, early in the morning. Aerators run the whole day since they are not able to monitor constantly. At night, a spotlight is turned on to provide heat in the water of the cage. Their only basis if the Dissolved Oxygen is low is when the Milkfish start to bubble. According to the farmers, bubbling is the action where the fish goes to the top of the water gasping for air. They conduct a change of water in the pond every Thursday of the week. Figure 2 shows the existing system based on the interview of the fishpond farm owners.

2.3 Description of the Proposed System

In the proposed system as shown in Fig. 3, pH level, temperature, dissolved oxygen, and turbidity each has a sensor used to continuously monitor their respective values that correlates with the water quality. These parameters have certain conditions depending on their optimum level.

For pH level, if the sensor monitored that the value of pH is less than 7, the system will release lime and alkaline fertilizer, and acid fertilizer, if greater than 8 to stabilize the pH level. In measuring the temperature of the milkfish pond, the proponents used a temperature sensor. When the temperature reaches below 24 °C, the heater will turn on, and when it reaches higher than to 30 °C, the aerators will also turn on.

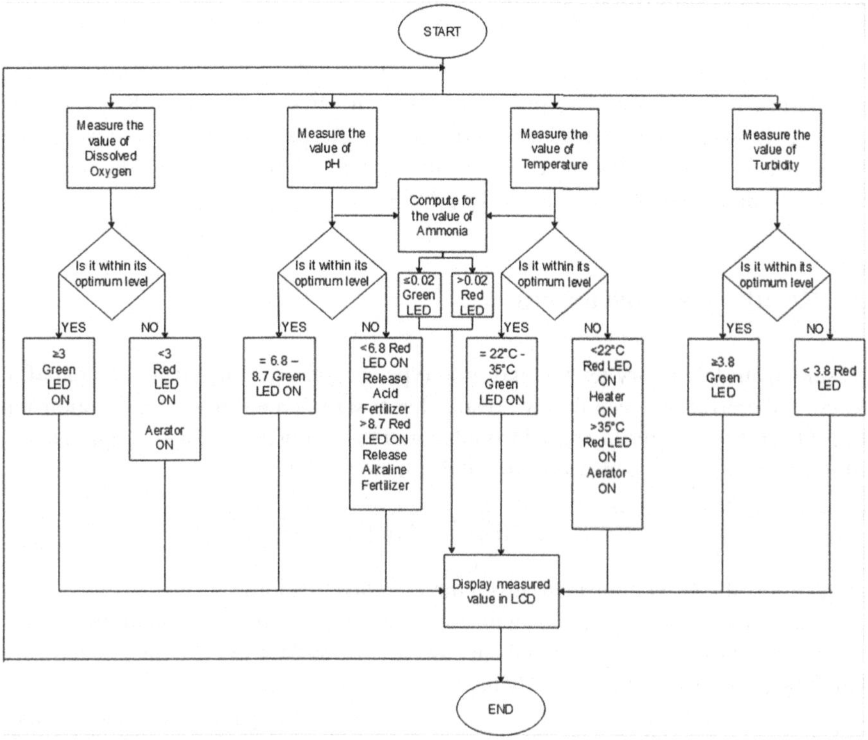

Fig. 3 Process flowchart of the proposed system

In measuring the cloudiness of water that can cause mud clay, the proponents used turbidity sensor. When the cloudiness of water reaches the minimum point of 0.5 NTU, the water filter will turn on. This will help to prevent forming of mud clays. Lastly, it is stated that when Dissolved Oxygen is consistently low and the water is supersaturated, i.e. above 300%, Gas Bubble disease may occur. The proponents used dissolved oxygen sensor to monitor the dissolved oxygen in the milkfish pond. If the dissolved oxygen is lower than 3 ppm, the aerators will turn on. This will help to mix the oxygen and water that will stabilize the condition of the milkfish pond in terms of dissolved oxygen.

To help the farmer see the current situation of their farms, the proponents used an LCD that will show the values of the ammonia, pH level, temperature, turbidity, and dissolved oxygen. LEDs were used to identify if the parameters are within optimum level.

2.4 *Materials and Design*

The primary components used in the prototype are the pH Meter Sensor Analog Kit, Temperature Sensor Waterproof DS18B20 kit, Dissolved Oxygen Kit, Turbidity Analog Sensor, Ammonia Sensor and Arduino Mega 2560 Rev3. For software, C++ was used in writing the codes for the system. Table 2 contains the materials used with their basic specifications (Fig. 4).

3 Results and Discussion

Results from laboratory testing and prototype sensors' readings were compared to know if the error between the two results has a significant difference. Relationship of NTU unit from turbidity based from laboratory testing and voltage output reading of the prototype was calibrated using the equation below.

$$y = -1120.4x^2 + 5742.3x–4352.9 \tag{1}$$

Where: y is the NTU equivalent given the x voltage response.

Since the fishpond is enclosed with concrete materials, deviation in turbidity level throughout the day of acquisition from 8:00 AM 12:00 MN with 1 h interval has very small ranging, from 3.85 to 3.89 Volts.

For the Ammonia sensor's reading, Eq. 2 was used to get the equivalent values of the output.

$$\left(1/\left(10\hat{}0.0901821 + 2729.92/(273.15 + B2)\right) - A2\right) + 1\,))\,(0.082) \tag{2}$$

Where: A2 – is the temperature value and B2 is the pH value. Summary of the comparison between the laboratory test and sensors measurement was analyzed as can be seen in Table 3.

Results of turbidity's reading in volts when converted to NTU using Eq. (1) above have undesirable results. This is normal due to the fact that the output response of the turbidity sensor is not linear. In the case of ammonia, equivalent reading was based on the function of temperature and pH reading equivalence. Comparison of laboratory results to sensor reading will have a biased result if this will be compared to laboratory results. From the reasons mentioned above, it is not advisable to compare the two reading to laboratory reading.

Since accuracy in sensor's reading is a significant factor in activating the controls to maintain the values at optimum level, it is necessary to know if the reading of the sensors are accurate. Data from 3 days' readings were analyzed, using standard deviation and mean difference. Results in Table 4 showed that the sensors' readings are accurate, with a very small standard deviation and mean difference values on all the sensors in 3 days' readings, with 16 hours duration using the interval of 1 hour data acquisition from 8:00 AM to 12 MN.

Table 2 Materials and specifications

Components	Specifications
pH meter sensor analog kit 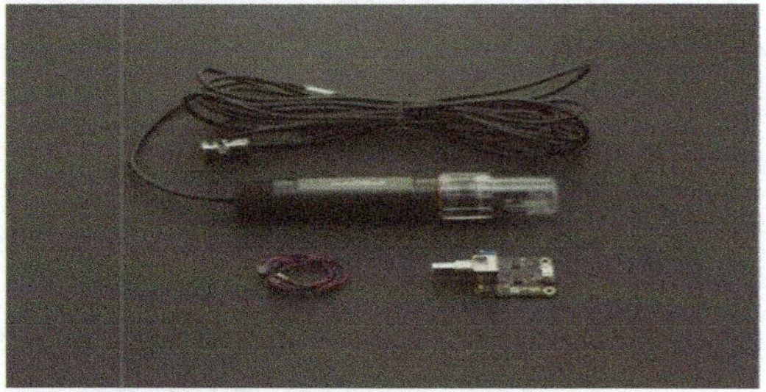	Module power: 5.00 V Module size: 43 mm × 32 mm (1.70″ × 1.26″) Measuring range: 0–14 PH Measuring temperature: 0–60 °C Accuracy: ± 0.1pH (25 °C) Response time: ≤ 1 min Industry pH electrode with BNC connector PH2.0 Interface (3 foot patch) Gain adjustment potentiometer Power indicator LED
Temperature sensor waterproof DS18B20 kit 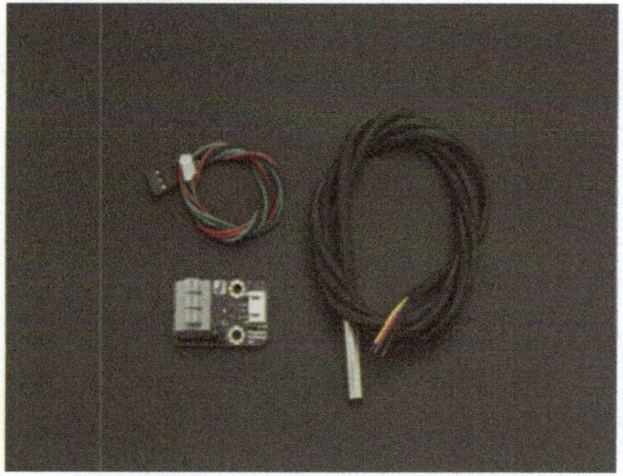	Operating voltage: 3.0~ 5.5 V ±0.5 °C accuracy from −10 °C to +85 °C Usable temperature range: −55 to 125 °C (−67 ° F to +257 ° F) To 12 bit selectable resolution Unique 64 bit ID burned into chip Multiple sensors can share one pin Temperature-limit alarm system Query time is less than 750 ms

(continued)

Table 2 (continued)

Components	Specifications
Dissolved oxygen kit 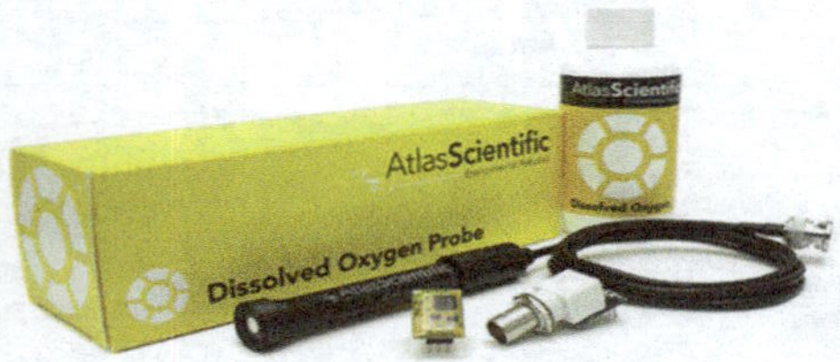	Range: 0–20 mg/L Probe body material: Epoxy and Noryl Max probe temperature: 50 °C Max probe PSI: 690 kPa (100PSI) Full range D.O. readings ±0.1 Accuracy within two significant figures (XX.XX mg/L) Temperature dependent or temperature independent readings Fresh water/saltwater/brackish water readings Conductivity-dependent or conductivity-independent readings Simple 5 s calibration Data output is in mg/L
Turbidity analog sensor 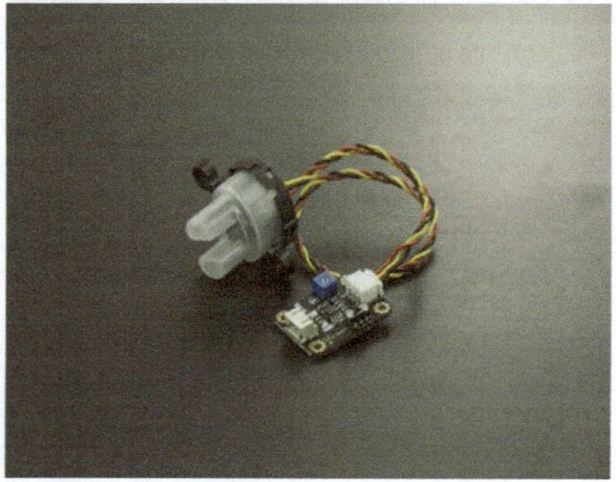	Operating voltage: 5 V DC Operating current: 40 mA (MAX) Insulation resistance: 100 M (min) Output method: analog Analog output: 0–4.5 V Digital output: High/low level signal (you can adjust the threshold value by adjusting the potentiometer) Operating temperature: −30 °C~80 °C Storage temperature: −10 °C~80 °C Weight: 30 g

Fig. 4 Prototype design

Table 3 Comparison of laboratory result and sensors readings, using percent error

Parameters	Laboratory result	Measured value from sensors	Percentage error
Ammonia	<0.01	0.0004 mg/L	N/A
Dissolved oxygen	8 mg/L	7.94 mg/L	0.75%
pH	7.61	7.41	2.63%
Temperature	20.2	20.52 C	1.58%
Turbidity	0.95 NTU	4.07 V	N/A

4 Conclusion

The system's prototype was able to perform the desired output, which is to measure the pH, Ammonia, Dissolved Oxygen, temperature, and turbidity. Results of sensors' readings were compared to the laboratory results through percent error and found that reading from dissolved oxygen, pH, and Temperature has less than 5% error. This result suggested that the system's prototype design is highly accurate. Standard deviation and mean difference were obtained from the series of testing to verify the accuracy of the sensors. Results of all sensors showed very small values of standard deviation and mean, suggesting that the sensors were accurate.

Table 4 Summary of testing results with statistical treatment of data

Day	Parameters	Mean	SE(x)	σ^2	σ	s^2	s
1	Ammonia	0.001123	0.000146	3.41E-07	0.00058384	3.62E-07	0.000602
	Dissolved oxygen	8.646471	0.025335	0.010270	0.1013405	0.01091	0.104459
	pH	7.323529	0.074434	0.088646	0.2977354	0.09419	0.306897
	Temp	24.31118	0.059981	0.057563	0.23992357	0.06116	0.24731
	Turbidity	3.867648	0.002641	0.000112	0.01058824	0.00012	0.010914
2	Ammonia	0.003191	0.000206	6.79E-07	0.0008239	7.21E-07	0.000849
	Dissolved oxygen	8.847647	0.020749	0.006889	0.08299748	0.00732	0.085552
	pH	7.853530	0.031482	0.015858	0.12592907	0.01685	0.129805
	Temp	24.46765	0.025574	0.010465	0.10229884	0.01112	0.105447
	Turbidity	4.344706	0.004298	0.000296	0.01719016	0.00031	0.017719
3	Ammonia	0.009971	0.000723	8.37E-06	0.00289333	8.89E-06	0.002982
	Dissolved oxygen	9.145294	0.027118	0.01177	0.10847161	0.0125	0.111810
	pH	8.372353	0.039691	0.025206	0.15876469	0.02678	0.163651
	Temp	24.79118	0.029166	0.013610	0.11666354	0.11666	0.12025
	Turbidity	3.977059	0.006793	0.000738	0.02717367	0.00078	0.028010

5 Recommendations

The overall performance of the system and designed prototype was satisfactory, but some recommendations will be stated in this section for future improvement and applications. During the designing process, the sensor should be calibrated properly. It is recommended to check whether the device has a linear or exponential output response. This is necessary to calibrate the output initially through hardware calibration. Such as creating a circuit design to correct the signal output based on the datasheet. Calibration should be done properly when dealing with device having nonlinear response, since a small change in sensor's reading will have a huge impact in the unit equivalence. It is also recommended to include security features in the system, backup power source, and interactive monitoring.

References

1. Lucy Towers. (2012). *The long and Short of Milkfish trading in the Philippines*. The Fish Site. Retrieved June 15, 2012, from https://thefishsite.com/articles/the-long-and-short-of-milkfishtrading-in-thephilippines
2. High value fish and milk fish production. (n.d.). *Domestic investment promotions service board of investments. Bureau of Fisheries and Aquatic Resources*. Retrieved from http://investphilippines.gov.ph/high-value-fish-and-milk-fish-production/

3. Beyond milkfish and tilapia: Future directions of Philippine aquaculture. (2017). Retrieved from http://www.nast.ph/index.php/downloads/category/118-39th-annual-scientific-meeting?download=415:plenary-3-4-beyond-bangus-and-tilapia-future-directions-of-philippine-aquaculture. National Academy of Science and Technology, Philippines.
4. Datta, S. (n.d.). *Management of water quality in intensive aquaculture*. Retrieved from https://www.researchgate.net/profile/Dr_Subhendu_Datta/publication/259175404_Management_of_Water_Quality_in_Intensive_Aquaculture/links/0c96052a1e592451fb000000/Management-of-Water-Quality-in-Intensive-Aquaculture.pdf
5. Francis-Floyd, R., Watson, C. (2006). *Ammonia*. The Fish Site. Retrieved August 28, 2006, from https://thefishsite.com/articles/ammonia
6. Watson, R., & Floyd, F. (n.d.). *Dissolved oxygen for fish production*. University of Florida Extension. Institute of Food and AgriculturalSciences. Retrieved from https://articles.extension.org/sites/default/files/w/3/33/Dissolved_Oxygen_for_Fish_Production.pdf.
7. Bhatnagar, A., & Devi, P. (2013). *Water quality guidelines for the management of pond fish culture*. Retrieved from http://www.ipublishing.co.in/ijesarticles/thirteen/articles/volthree/EIJES31197.pdf
8. Managing High pH in Freshwater Ponds. (2013). Retrieved, September 30, 2013. From https://thefishsite.com/articles/managing-high-ph-in-freshwater-ponds
9. Turbidity, Total Suspended Solids & Water Clarity (n.d.). Retrieved from http://www.fondriest.com/environmental-measurements/parameters/water-quality/turbidity-total-suspended-solids-water-clarity
10. Mayer, E. (2012). *Monitoring pond water quality to improve production*. The fish site. Retrieved June 15, 2012, from https://thefishsite.com/articles/monitoring-pond-water-qualityto- improveproduction.

Optimization of Nonlinear Temperature Gradient on Eigenfrequency Using Genetic Algorithm for Reinforced Concrete Bridge Structural Health

Ronnie S. Concepcion II, Lorena C. Ilagan, and Ira C. Valenzuela

1 Introduction

Even with the advancement of mechanical and structural tools and systematic engineering of construction materials, the civil engineering sector faces problems on structural health integrity due to climate change, which demands safety. A bridge is considered the most important civil infrastructure due to the fact that it plays a vital role in the economic affluence of the nation and merely provides road to disconnected locations [1]. It has been evident that changes in environmental condition, particularly in temperature, results in much observable variations in modal vibration properties, namely, frequencies, mode shapes, and damping of civil infrastructures [2, 3]. Bridge and environment temperatures are the two distinct temperatures that are imperatively interacting on the output modal response of a bridge and has an undeniable effect. Thermal gradient is present along the deck of the bridge [4]. With the bridge being subjected to thorough thermal loading,

R. S. Concepcion II (✉)
Department of Electronics Engineering, University of Perpetual Help System DALTA, Las Piñas City, Metro Manila, Philippines

School of Graduate Studies, Mapua University, Intramuros, Manila, Philippines
e-mail: concepcionriis@gmail.com

L. C. Ilagan
Department of Electronics Engineering, University of Perpetual Help System DALTA, Las Piñas City, Metro Manila, Philippines

Electronics and Communications Engineering Department, De La Salle University, Manila, Philippines

I. C. Valenzuela
Electronics and Communications Engineering Department, De La Salle University, Manila, Philippines

A. Beltran Jr. et al. (eds.), *World Congress on Engineering and Technology; Innovation and its Sustainability 2018*, EAI/Springer Innovations in Communication and Computing, https://doi.org/10.1007/978-3-030-20904-9_11

whether it is linear, nonlinear, or uniform, creates stresses with same magnitude of the live loads, alters the eigenfrequency of the bridge, and results in expansion and contraction of material [5]. Eigenfrequency is the natural oscillatory frequency of a system without any excitations and induced forces [6]. It is the infrastructure characteristic frequency that does not require full transient test. In order to quantify eigenfrequency, vibration value must be retained and all parameter effects must be eliminated.

There are numerous researches that focus on correlating the physical properties of the bridge and the observable excitations, such as vehicle loading, with the resulting modal frequency [7–9]. To maximize the effectivity of the design and the functionality of the system, optimization tool can be used [10–12]. There are studies that used genetic algorithm for developing estimations on structural damage using dynamic parameters data, and considers deformability and static parameters along the beams of the bridge [13, 14]. However, these kinds of approaches require multiple parameters to contextualize the response and results to complex analysis.

This study aims to determine the suitability of hybrid principal component analysis and genetic algorithm approach in finding the optimum condition for a reinforced concrete bridge. Steel bridge is often studied because of its high sensitivity to temperature. In this study, eigenfrequency of reinforced concrete bridge sensitivity to temperature was analyzed. The effect of bridge temperature and environment temperature was sufficiently compensated to retain the integral system characteristic using principal component analysis. These temperatures were optimized using genetic algorithm. The knowledge on these two thermal parameters will help civil and structural engineers in improving their structural integrity analysis and design.

2 System Architecture

The developed conceptual process flow is shown in Fig. 1. Principal component analysis (PCA) was used in this study for its efficiency not only in multivariable reduction but also in compensating parametrical effect, while reducing the majority of its original characteristic [15]. Genetic algorithm (GA) was used for its simplicity and capability in finding an optimal solution to a specific problem [16–18]. This metaheuristic evolutionary genetic algorithm is a higher-level procedure designed to find approximate solutions to difficult-to-solve problems, mainly with imperfect and incomplete information, making it appropriate in determining the best condition of the bridge.

In this study, bridge and environment temperatures are the variables that will be optimized. The PCA-reconstructed data is the temperature-compensated data that will be initialized as input to GA. Initial random population will be created. A sequence of new populations will then be created based on the fitness score, as the system selects 50 fittest individuals. The individuals in each generation are used to create the new set of population. Then, recombination process will take place by combining the vector entries of a pair of parents. Mutation will be applied to verify

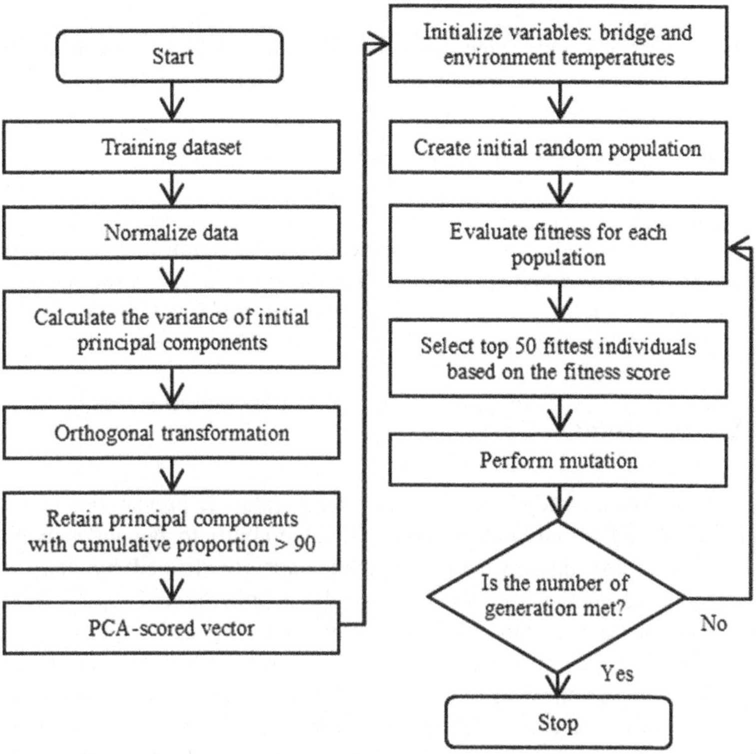

Fig. 1 System process flow using hybrid PCA-GA

if the solution is global, by making random changes to a single parent. The system will permanently stop and output the optimum condition for a bridge when iterations had reached the required number of generation.

The algorithm takes bridge and environment temperatures from unobservable environmental parameters, where the bridge is located as input to the system with artificial excitations using subwoofer speaker located at the top of the deck. The system is designed to provide the optimum condition of the bridge with no damage. The output of the system is the best bridge and environment temperature levels.

2.1 *Principal Components*

Principal component analysis (PCA) serves as pretreatment algorithm for the input vectors. It is a mathematical combination of pattern recognition and supervised machine learning that emphasizes variation in data yields' strong patterns. Temperature is the most important unobservable environmental parameter that severely affects the stability of a bridge. PCA is based on cumulative proportion (CP) criterion that suppresses most of the dependent parameters, temperatures to the modal frequency. This yields eigenfrequency. As shown in Eq. 1, λ is the eigenvalue, E is the nth number of eigenvalue and P is the number of variables.

$$\mathrm{CP} = \frac{\sum_{k=1}^{E} \lambda_k}{\sum_{P=1}^{P} \lambda_P} \tag{1}$$

Principal component (PC) is the eigenvector of the eigenvalues of data matrix, and is basically the number of dependent parameters retained in the model that is the temperatures. A total of 2160 data rows were inputted to a PCA subsystem.

2.2 *Chromosomes*

A reinforced concrete bridge requires stable temperature to prevent drastic material degradation. With thermal gradient loadings, vulnerability to damage is incurred.

The chromosome data structure is mathematically defined by N_{ind} x L_{ind}, where N_{ind} is the number of individuals in the population and L_{ind} is the length of the genotypic representation of each individual. Figure 2 depicts the genetic components of a single gene. The unit used for measuring both temperatures is °C.

The initial generation was set to contain 50 chromosomes using random generation from the entire range of search space. It comprises varying genotypic bases with equal length and nonuniform distance structure. Observed bridge and environment temperatures during vibration excitation were used as input.

2.3 *Fitness Function*

Fitness function is a figure of merit used to evaluate the effectiveness of each chromosome in a phenotype. A phenotype is considered a decision variable in which each genotype interacts with another. In this study, a two-temperature eigenfrequency model has been used, as shown in Eq. 2. The V is the z-axis vibration relating to eigenfrequency, *bridgeTemp* is the temperature along the deck of the bridge, and *enviTemp* is the ambient temperature affecting the bridge. This model was constructed using the multiple linear regression (MLR) and the temperature-

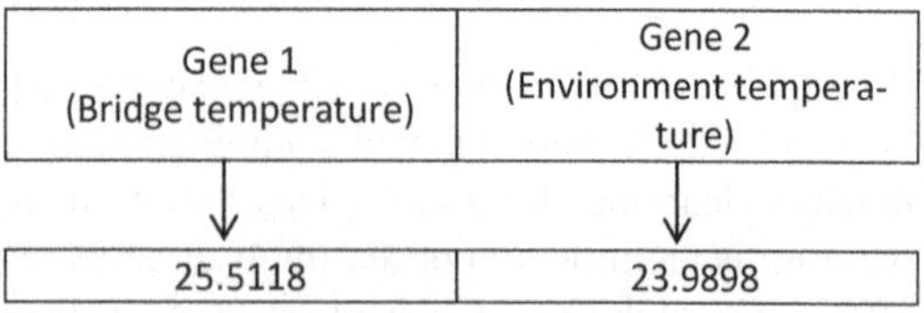

Fig. 2 Sample chromosome data structure with its genotypic representation

compensated vibration response of the PCA subsystem. This two-temperature model was transformed to its fitness function, as shown in Eq. 3. The x_1 resembles bridge temperature and x_2 the environment temperature.

$$V = 1.1676 - 0.01523\ (\text{bridgeTemp}) + 0.00774\ (\text{enviTemp}) \tag{2}$$

$$f\,(x_1, x_2) = 1.1676 - 0.01523 * x_1 + 0.00774 * x_2 \tag{3}$$

2.4 Selection

The best chromosomes from the entire population are selected after the fitness value was computed to breed a new generation. It is basically selecting individuals to be contributing parents for the next generation. There are various selection methods, including stochastic uniform, remainder, uniform, roulette wheel, rank, tournament, Boltzman selection, reward-based selection, fitness proportionate, and steady-state selection that rate the fitness of each solution and select the best solutions [19–21]. In this study, the tournament selection method was used due to the selection pressure provided to each individual. This selection method runs several tournaments among chromosomes that were randomly chosen from the population. The one with the best fitness is considered the winner of each tournament and is selected for the next process that is a crossover. The selection pressure is the probability of likelihood of each chromosome to participate in the tournament. As each genotype differs from the others, selection pressure is materialized, thus it makes GA optimize the subsequent genes. With this, the optimal solution is based on the efficiency of GA using this selection method.

For nonlinear constraint algorithm, a penalty is added in selecting the fittest chromosome. Chromosomes that are not qualified will be discarded and all the fittest chromosomes, including the parents and offspring, form the next generation by undergoing reproduction.

2.5 Crossover

Crossover combines the genetic information of two parents to form children for the next generation, thus, otherwise known as genetic recombination [16, 17]. It assorts the characteristics from two parents [22]. The results are newly partial generated genes or solutions that must undergo mutation before being added to the actual population. There are different crossover techniques such as constraint dependent, scattered, single point, two point, intermediate, heuristic, and arithmetic.

Single-point crossover was chosen for this study. Figure 3 shows that either of the genes will be replaced by the algorithm.

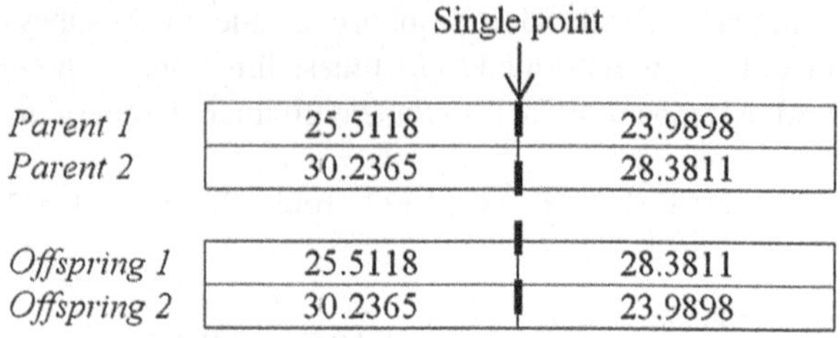

Fig. 3 Sample of genotypic representation of single-point crossover

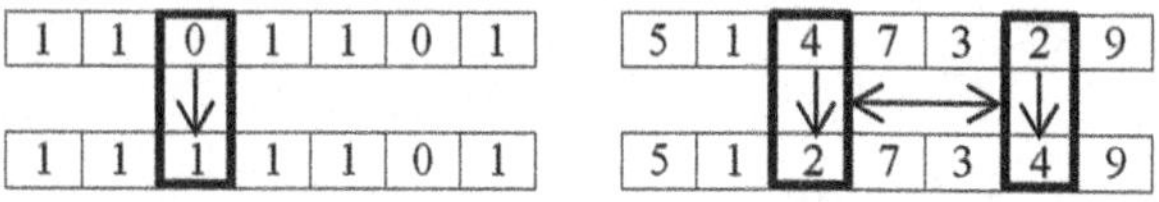

Fig. 4 Sample 7-bit string mutation and integer mutation

2.6 Mutation

Mutation introduces random genetic changes to individual parents to form children, thus, it maintains the genetic diversity from one genetic chromosome generation to the next. Local minima are the tendency of two or more chromosomes becoming alike, and it is avoided by the mutation process [18]. There are different mutation techniques such as boundary, bit string, constraint dependent, flip bit, Gaussian, nonuniform, uniform, adaptive feasible, and shrink (Fig. 4).

Uniform mutation was chosen for this study, as it is effective for integer and float genes. The algorithm replaces the selected fraction of vector entries with the uniformly selected random value from the search space.

3 Optimization Using Genetic Algorithm

In optimizing the two-temperature eigenfrequency model of a bridge's structural health using genetic algorithm, some parameters have fixed values and some have varying values. The parameters with fixed values are the two optimization variables, bridge and environment temperatures, the number of generation that is equal to 100, and the initial population size of 50. Contrariwise, the parameters with varying values are the three genetic operators used to generate and optimize new generations, which are selection rate, crossover rate, and mutation rate.

In Figs. 5 and 6, bridge and environment temperatures have varying effects on the bridge frequency. As the bridge temperature increase, the bridge frequency also increases. Majority of the temperature-frequency data lies on the 23–32 °C range. It is also considerable that, as the value of temperature becomes lower, the dominant temperature-frequency data lies at 1–1.7 Hz. Thus, nonlinear thermal loadings affect the eigenfrequency of the bridge.

Figure 7 depicts the interaction of the two optimization variables, bridge and environment temperatures, with the bridge frequency. It is evident that, as

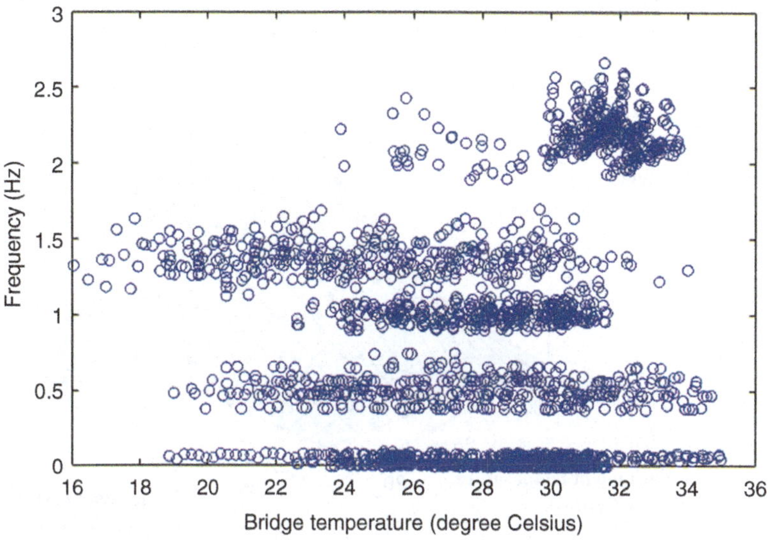

Fig. 5 Effect of bridge temperature on bridge frequency

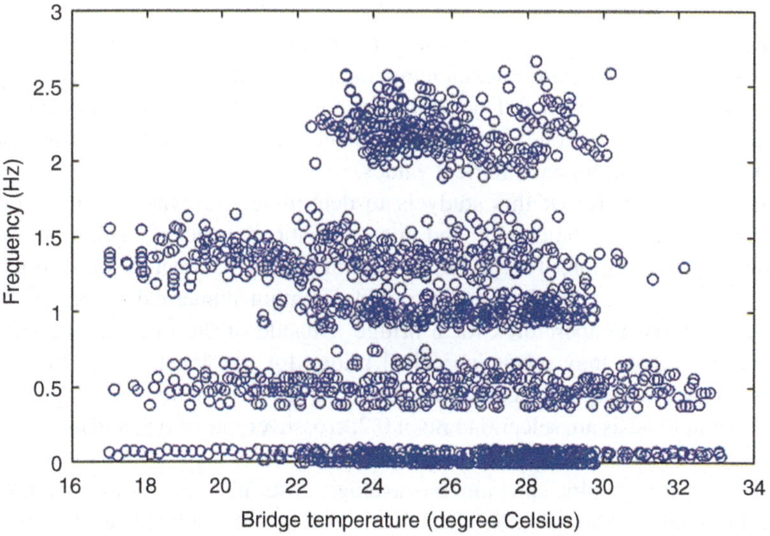

Fig. 6 Effect of environment temperature on bridge frequency

temperatures linearly increase and cause thermal gradient along the deck of the bridge, the bridge frequency also increases. Moreover, it is noticeable that starting at 25 °C for the bridge temperature, the corresponding frequency response is almost constant to its maximum level. On the other hand, the bridge frequency peaks at 25 °C for the environment temperature. Decreasing the environmental

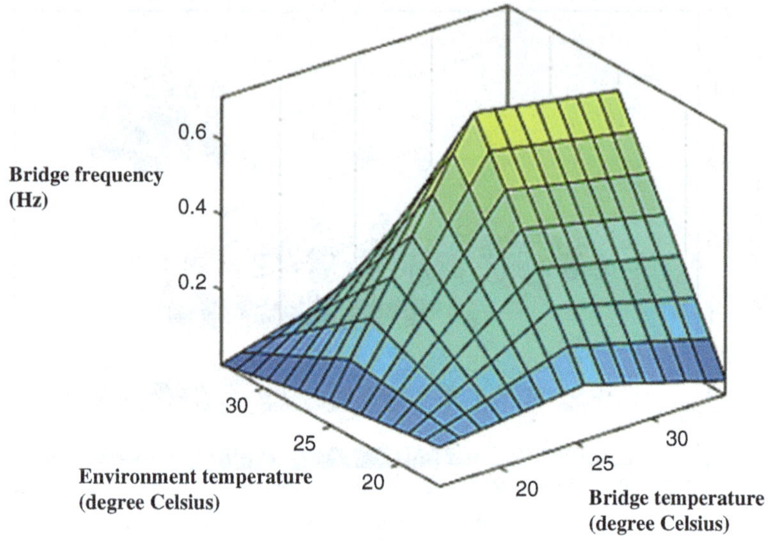

Fig. 7 Interaction of bridge and environment temperatures with the bridge frequency

temperature from 25 °C level results in linearly decreasing bridge frequency. Likewise, increasing environmental temperature from 25 °C level results in linearly decreasing bridge frequency. Due to the nonpatterned and varying effects of these temperature parameters on bridge eigenfrequency, genetic algorithm is necessary to determine the optimum parametrical values.

The main objective of this study is to determine and evaluate the suitability of hybrid principal component and metaheuristic genetic algorithm in finding the optimum conditions for bridge structural health parameters on temperature basis. GA was critically designed to yield optimum numerical values of bridge and environment temperatures for a bridge. Outside of the tolerable values from the optimal values may result in critical bridge for the worst case. After several explorations, the design that gives the optimal values for structural health parameters on temperature basis are selection rate of 0.2, crossover rate of 0.9, and mutation rate of 0.01.

The best fitness plot showing the average costs per population and the best individuals are shown in Fig, X. The value for the best individual in each population is 0.961830, and the average population fitness value is 0.963103 (Fig. 8).

Table 1 shows the optimum values obtained using genetic algorithm for bridge and environment temperatures. Outside of these temperature values, the bridge structural health may be labeled as needing rehabilitation or critical, in the worst case. Thus, temperature ratios with large differences, compared to the generated optimized values, renders possible breakage or damage in the physical structure of the bridge.

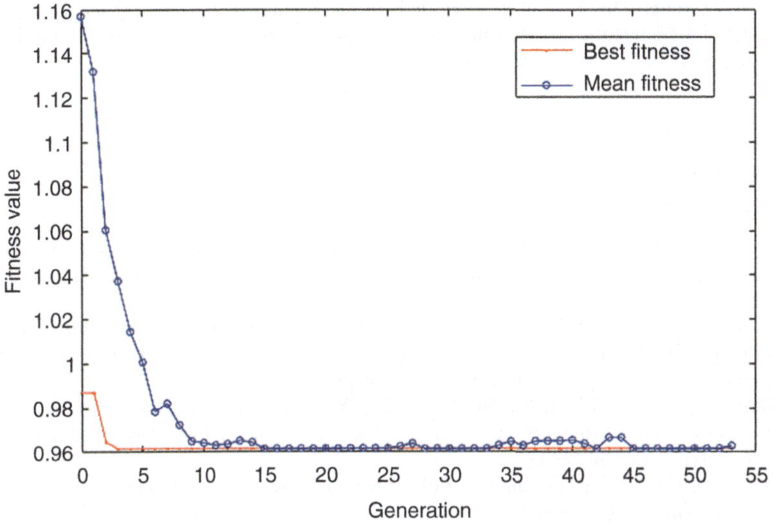

Fig. 8 Best fitness plot

Table 1 Results of genetic algorithm

Structural health parameter	Optimum value
Bridge temperature	9.578 °C
Environment temperature	−8.571 °C

4 Conclusion and Future Work

This study presented the suitability of applying the hybrid principal component analysis and metaheuristic genetic algorithm approach in determining the optimum values of bridge and environment temperatures. Their optimization factors are the most essential, unobservable environmental effects that greatly affect the eigenfrequency of the bridge. The graphs show varying effects of the optimization parameters on bridge frequency, rooting the need of genetic algorithm in this study. Critical design of genetic algorithm was done and the best fitness plot was obtained. The genetic algorithm showed the optimum conditions for good condition of a bridge, needing a bridge temperature of 9.578 °C and environment temperature of 8.571 °C. It is not necessary to have exactly these temperature levels, since bridges located in different places having diverse climates are subject to variations, but at least to have almost the same temperature difference with the optimized values may result in good bridge structural health condition.

Future works include using particle swarm optimization and simulated annealing technique to verify the most effective metaheuristic optimization method for this kind of data.

Acknowledgments This project is supported by the Engineering Research and Development Technology program of the Department of Science and Technology, and the Faculty of Engineering of the University of Perpetual Help System DALTA, Las Piñas Campus.

References

1. Shi, A., & Yu, X. (2012). Structural damage detection using artificial neural networks and wavelet transform. *IEEE Computational Intelligence for Measurement Systems and Applications*.
2. Zhang, H., Gui, J., Xie, Y., & Sun, Y. (2013). Environmental effect removal based structural health monitoring in the internet of things. *IEEE International Conference on Innovative Mobile and Internet Services in Ubiquitous Computing*.
3. Xia, Y., Chen, B., Weng, S., Ni, Y., & Xu, Y. (2012). *Temperature effect on vibration properties of civil structures: A literature review and case studies*. Hong Kong: The Hong Kong Polytechnic University.
4. Childs, D. (2016). The response of full bridge strain sensors to thermal variations. *United Kingdom National Quantum*.
5. El-Tayeb, E., El-Metwally, S., Askar, H., & Yousef, A. (2015). *Thermal analysis of reinforced concrete beams and frames*. Housing and Building National Research Center.
6. Peeters, B., Maeck, J., & Roeck, G. (2015). Vibration-based damage detection in civil engineering: Excitation sources and temperature effects. *Smart Materials and Structures*.
7. Pakzad, S., & Fenves, G. (2009). Statistical analysis of vibration modes of a suspension bridge using spatially dense wireless sensor network. *American Society of Civil Engineer*.
8. Cook, W. (2014). *Bridge failure rates, consequences, and predictive trends*. Utah State University Digital Commons.
9. Lynch, J. (2002). *Decentralization of wireless monitoring and control technologies for smart civil structures. Ph. D. Thesis*. Department of Civil and Environmental Engineering, Standford University.
10. Bureerat, S., & Cooper, J. (1998). Evolutionary methods for the optimizations of engineering systems. *IEEE Colloquium on Optimization and Control: Methods and Application*.
11. Chaturvedi, S., Pragya, P., & Verma, D. (2015). Comparative analysis of particle swarm optimization, genetic algorithm and krill herd algorithm. *IEEE International Conference on Computer, Communication and Control*.
12. Fidrysiak, Y., Przewozniczek, M. (n.d). Towards finding an effective way of discrete problems solving: The particle swarm optimization and linkage learning techniques hybridization. IEEE.
13. Wei-jian, Y., & Xia, L. (2001). *Damage diagnosis of structures by genetic algorithms*. Hunan University.
14. Sgmabi, L., & Bontempi, F. (2005). A genetic algorithm approach for performance evaluation of long span suspension bridges. *International Conference on Civil, Structural and Environmental Engineering Computing*.
15. Concepcion, R., Cruz, F., & Uy, F. (2017). Supervised PCA for temperature compensation of vibration data of concrete bridge. *Journal of Telecommunication, Electronics and Computer Engineering*.
16. Haupt, R., & Haupt, S. (2004). *Practical genetic algorithm*. New Jersey: John Wiley and Sons, Inc.
17. Mitchell, M. (1999). *An introduction to genetic algorithms*. London: MIT Press.
18. Sivanamdam, S., & Deepa, S. (2008). *Introduction to genetic algorithms*. New York: Springer.
19. Kuncheva, L., & Jain, L. (2000). Designing classifier fusion system by genetic algorithm. *IEEE Transactions on Evolutionary Computation*.

20. Bedruz, R. (2016). Philippine vehicle plate recognition using image thresholding and genetic algorithm. *IEEE Region 10 Conference*.
21. Uy, A., et al. (2015). Machine vision for traffic violation detection system using genetic algorithm. *International Conference on Humanoid, Nanotechnology, Information Technology, Communication and Control, Environment and Management*.
22. Baldovino, R., et al. (2017). GA optimization of coconut sugar cooking process: A preliminary study using stochastic universal sampling (sus) technique. *ICCAE*.

Alertness and Mental Fatigue Classification Using Computational Intelligence in an Electrocardiography and Electromyography System with Off-Body Area Network

Ronnie S. Concepcion II, Jommel S. Manalo, Ave Jianne D. Garcia, Rhaniel A. Legaspi, Jun Angelo Prestousa, Gio Paolo C. Pascual, Junco S. Firmalino, and Lorena C. Ilagan

1 Introduction

Human fatigue is a biological occurrence of decrease in readiness to execute certain activity that is not only associated with emotion, mental, and physiological aspects but also associates with tiredness and drowsiness. It is classified into two, namely, physical or muscular fatigue and mental fatigue. The former results from strenuous activities, and the latter has still unknown cause. Mental fatigue is a refined and indirect feeling, resulting in an unwillingness to perform any kind of activity. It is characterized by tiredness. Whereas drowsiness is the hardship of being awake. Both of these physiological cases are characterized by heart rate variability (HRV) and muscle activities [1]. HRV is frequently used as a quantitative marker to depict activity of autonomic nervous system (ANS) related to mental fatigue due to accumulated stress [2]. Redundant feeling of mental fatigue may act as symptom to more serious health problems, which must be diagnosed.

R. S. Concepcion II (✉)
Department of Electronics Engineering, University of Perpetual Help System DALTA, Las Piñas City, Metro Manila, Philippines

School of Graduate Studies, Mapua University, Intramuros, Manila, Philippines
e-mail: concepcionriis@gmail.com

J. S. Manalo · A. J. D. Garcia · R. A. Legaspi · J. A. Prestousa · G. P. C. Pascual · J. S. Firmalino
Department of Electronics Engineering, University of Perpetual Help System DALTA, Las Piñas City, Philippines

L. C. Ilagan
Department of Electronics Engineering, University of Perpetual Help System DALTA, Las Piñas City, Metro Manila, Philippines

Electronics and Communications Engineering Department, De La Salle University, Manila, Philippines

A. Beltran Jr. et al. (eds.), *World Congress on Engineering and Technology; Innovation and its Sustainability 2018*, EAI/Springer Innovations in Communication and Computing, https://doi.org/10.1007/978-3-030-20904-9_12

The fundamental signal from these physiological signals is the EEG signal, which is directly measured using scalp tapes and electrodes. In that way, clear vision of human brain activities can be recorded [3]. Various studies were done in assessing and detecting mental fatigue, using computer vision technology. It is an image-based system using physiological observations of visual behaviors, such as eye blink rate, eyelid movement, and the degree of eye opening or the percentage of eye closure (PERCLOS), the frequency of nodding, head angle, and the shoulder posture [4–8]. Physiological data such as high-resolution EEG signals, heart rate (HR), skin conductance, oximetry pulse, and eye blink rate (EBR) were also used to correlate metal fatigue and drowsiness [9, 10].

To continue this line of research, this study explores the use of computational intelligence techniques, in improving the accuracy of mental fatigue and alertness classification based on neurophysiological signals of electrocardiogram and electromyogram.

2 System Architecture

The conceptual flow of the system is shown in Fig. 1, in which each part complies with the objective of the study. The development of body access network enables the mote to communicate with the laptop, thus, completing a wireless communication system for data transmission. The implementation of soft algorithms verifies which of it generates the best detection of mental fatigue with minimum error.

2.1 Development of Body Area Network

Body area network (BAN) is otherwise known as body access network, which can provide health care to a disabled person, elderly, and even to persons with active lifestyle, who requires monitoring of their vital signs. It offers real-time sensing, data processing of valuable information, and transmission of physiological data for monitoring and analysis [11, 12]. BAN is categorized into three, namely, on-body surface (OBS) or wearable, in-body surface (IBS) or implanted, and off-body (OB) that depends on the placement of the sensor in the human body [13].

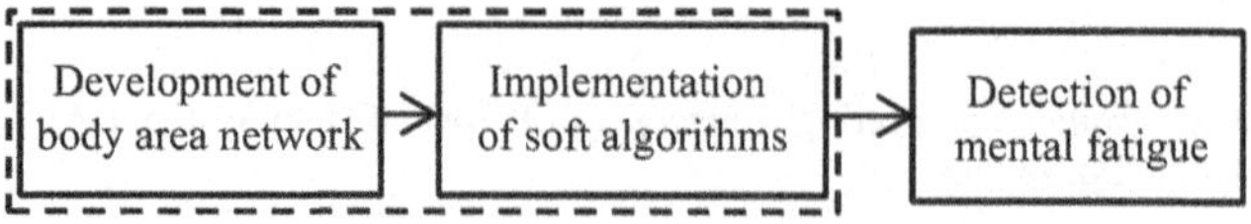

Fig. 1 Research architecture

Integration of Sensors

In this study, BAN architecture is virtually divided into data acquisition (DA), data processing (DP), and data transmission (DT) modules, as shown in Fig. 2. DA is composed of one thumb-sized heart rate sensor for ECG and one muscle contraction sensor for EMG. Mean square distance error (MSDE) was minimized during sensor calibration to provide higher accuracy and precision for testing and utilization. Mahalanobis distance (MD) was used to determine the gaussianity of measured physiological data. These sensors are directly plugged in to the microcontroller, with one 2-kilo-Ohm resistor for protection from surge current.

Integration of Bluetooth Module

HC-05 Bluetooth module was used as the DT component for the developed mote. It was configured as master and slave, with built-in antenna that covers up to 10 meters, making it a suitable solution for wireless communication in short distances, in which monitoring of neurophysiological data manifests.

Development of BAN Architecture

For any BAN, it requires at least one mote to communicate with another mote, wireless machine, or computer. A more complex BAN can utilize more than one mote intercommunicating in real time.

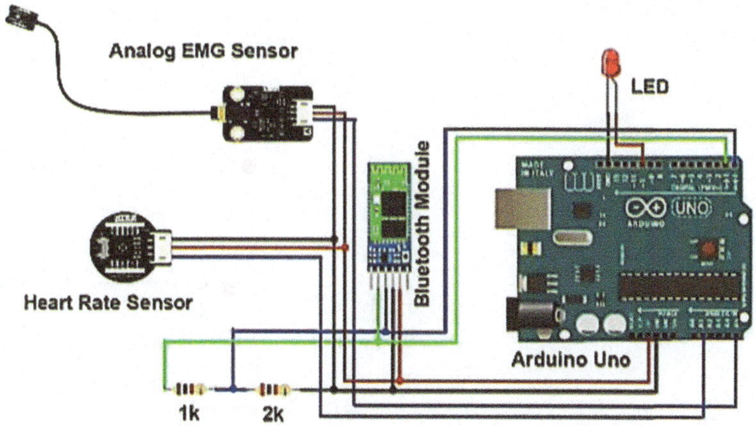

Fig. 2 Instrumentation diagram of mote

In this study, Arduino Uno microcontroller was used as the DP module. It has 14 digital input/output (I/O) pins, 6 analog input pins, and a 16 MHz quartz crystal for oscillation. C language was embedded in the microcontroller to control the operations of the network, including when to collect data and when to transmit to the computer for analysis. A data-centric BAN was constructed, using one mote that has the integration of ECG and EMG sensors, the Arduino microcontroller and the Bluetooth module, as shown in Fig. 3.

Development of Off-Body Area Network

The noninvasive collection of neurophysiological data of heart rate and muscle contraction utilized the DA module of the constructed BAN with straps tied on the upper arm and wrist, as shown in Fig. 4. Samples were collected from 20 university students of varying age, gender, and duration of sleep hours. It is restricted for the samplers to create unnecessary movements during data acquisition, for it might yield unreliable data. They were recommended to sit on a comfortable bench for 20 minutes with BAN data rate of 1070 Hz.

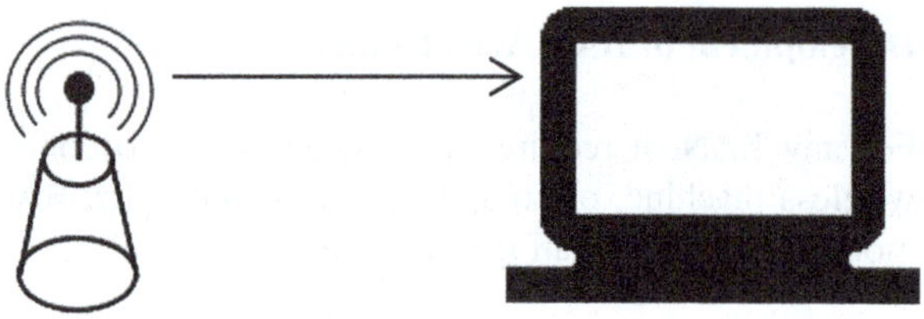

Fig. 3 BAN diagram

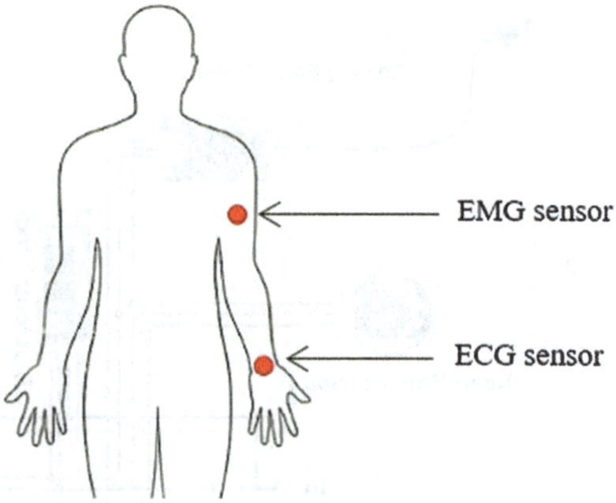

Fig. 4 Off-BAN sensor placement

2.2 *Implementation of Computational Intelligence*

Artificial Neural Network

Artificial neural network (ANN) is a computational biology that enables pattern recognition, classification, and machine learning. It is considered as an extreme learning machine. Neural network combines artificial neurons in processing information. It has inputs like synapses that are multiplied by the weights of the input signal and processed by a mathematical function to determine the formation of the neuron.

The development of ANN is composed of three parts: design of neural network, training of neural network, and testing of neural network.

The objective of the training process is to generate the target output for varying inputs. A three-layer feedforward neural network, with 10 artificial neurons residing at the hidden layer, was constructed in this study, as shown in Fig. 5. It is possible using backpropagation algorithm. In this case, signals from each neuron are being sent layer to layer, while the errors are propagated backward to affect another set of input signals. The purpose of the testing process is to verify the accuracy of the trained network.

Aside from tangential function, the sigmoidal function, as mathematically depicted in Eq. 1, is one of the most common activation functions for backprop-

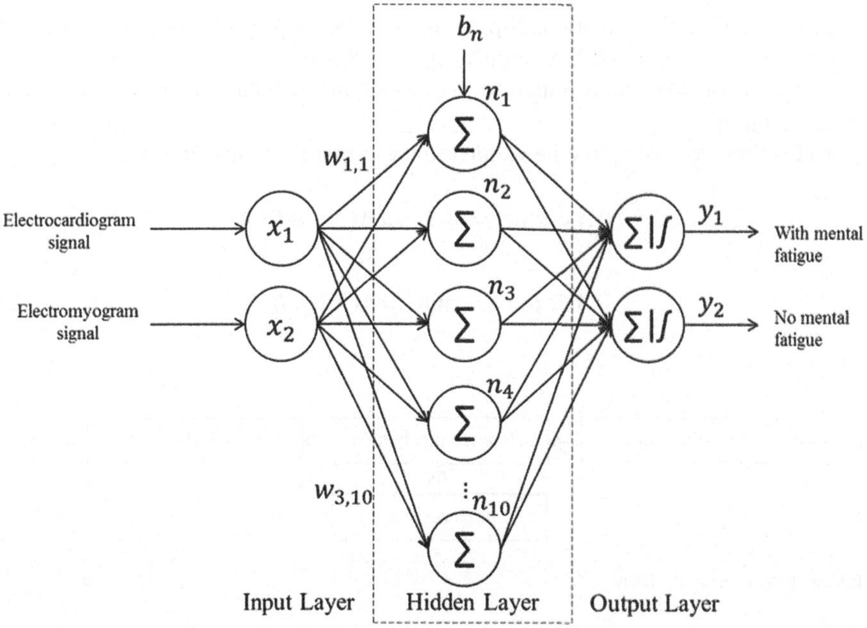

Fig. 5 Three-layer feedforward neural network with ten hidden artificial neurons

agation algorithm. It is the weighted sum of inputs xi multiplied by their respective weights w_{ji}. Eq. 2 defines the error function of a certain neuron that merely affects its output signal.

$$A_j(x, y) = \sum_{i=0}^{n} x_i w_{ji} \quad (1)$$

$$E_j(x, w, d) = \left(O_j(x, w) - d_j\right)^2 \quad (2)$$

The MATLAB nntool was used to initialize the number of artificial neurons, hidden layers, weights, and biases. Dual binary digit final grade representation was set for classified outputs.

Fuzzy Logic System

Fuzzy logic is a rule-based system, with arbitrary accuracy that is supported from a linguistic set of rules, for nonlinear mapping of scalar or vector crisp inputs and resulting in crisp outputs [14]. It focuses on how to provide the best classification base on the fuzzy rule and inference mechanism. As shown in Fig. 6, fuzzy logic system is divided into four main parts:

1. Fuzzification: converts crisp inputs to appropriate fuzzy values that can be easily contrasted with the fuzzy linguistic rules.
2. Fuzzy rule-based evaluation: clips all the membership functions as fuzzy knowledge on how to classify the inputs using Eqs. 3 and 4.
3. Inference mechanism: evaluations as to which membership rule is relevant to the fuzzy input.
4. Defuzzification: converts the yielded fuzzy output from inference to crisp value.

$$\mu A \cup B(x) = \max\ \{\mu A(X), \mu B(X)\} \quad (3)$$

$$\mu A \cap B(x) = \min\ \{\mu A(X), \mu B(X)\} \quad (4)$$

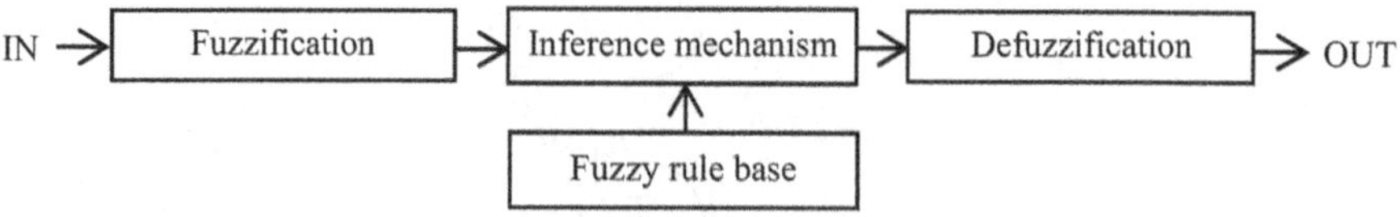

Fig. 6 Fuzzy logic structure

In this study, the fuzzy logic system for classifying alertness and mental fatigue consists of two inputs from heart rate and muscle contraction sensors. Based on these inputs, an output signal for classification was inherited. The description for membership function for heart rate is 0–200 beats per minute, and 0–800 millivolts for muscle contraction. The heart rate and muscle contraction inputs have three and two Gaussian membership functions, respectively. The output has two membership functions for activity level, namely, "alert" and "mental fatigue" (Figs. 7 and 8).

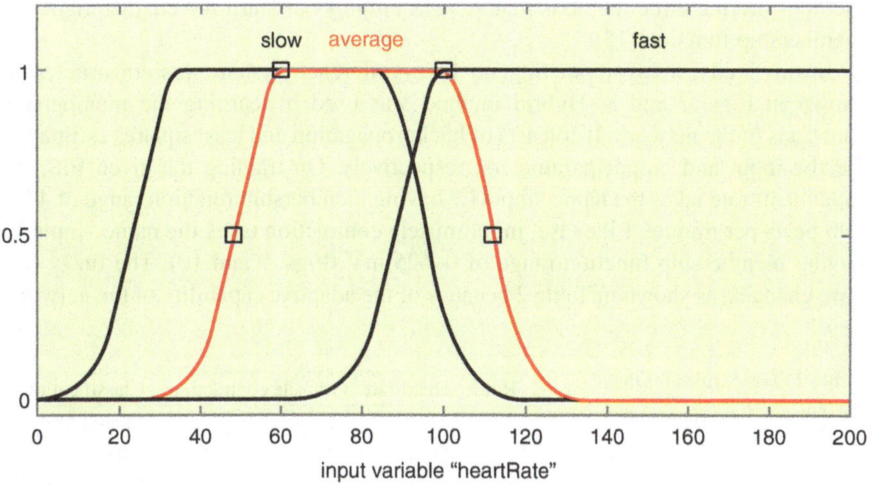

Fig. 7 Heart rate membership functions

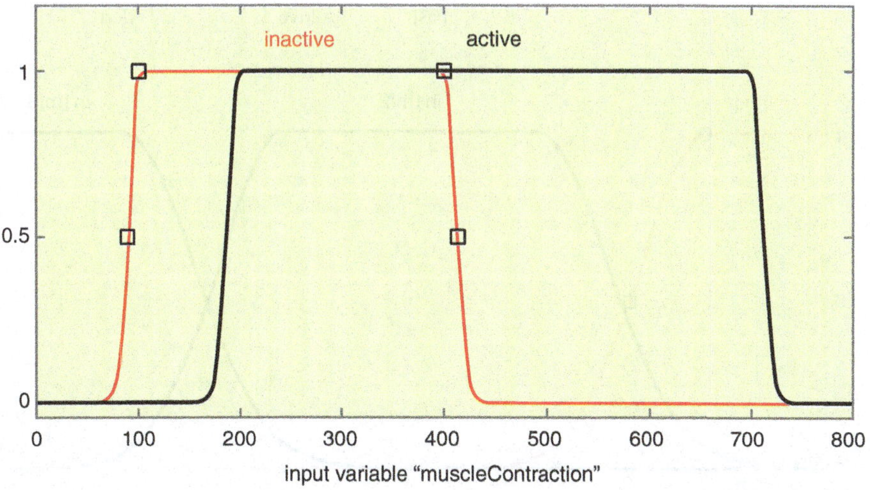

Fig. 8 Muscle contraction membership functions

The fuzzy rule base for the design is shown in Table 1. If heart rate is slow and the muscle contraction is inactive, then the person is assessed as having mental fatigue.

Adaptive Neuro-Fuzzy Inference System

Adaptive neuro-fuzzy inference system (ANFIS) is the combined learning network of ANN and fuzzy inference system (FIS) that is suitable in adaptive and optimization techniques. Backpropagation, hybrid method, and subtractive clustering methods are the three methods that ANFIS employs to learn the characteristics of membership function [15].

In this study, a two-input Sugeno fuzzy inference system was constructed as shown in Figs. 7 and 8. Hybrid method was used in learning the membership functions of the network. It consists of backpropagation and least squares estimation for the input and output parameters, respectively. On training the given FIS, the input heart rate takes the name "input1," having membership function range of 40 to 190 beats per minute. Likewise, input muscle contraction takes the name "input2," having membership function range of 0–625 mV (Figs. 9 and 10). The fuzzy rule base changes, as shown in Table 2 because of the adaptive capability of the network.

Table 1 Fuzzy rules for the design

Rule	Heart rate	Muscle contraction	Classification
1	Slow	Inactive	Mental fatigue
2	Slow	Active	Alert
3	Average	Inactive	Mental fatigue
4	Average	Active	Alert
5	Fast	Inactive	Mental fatigue
6	Fast	Active	Alert

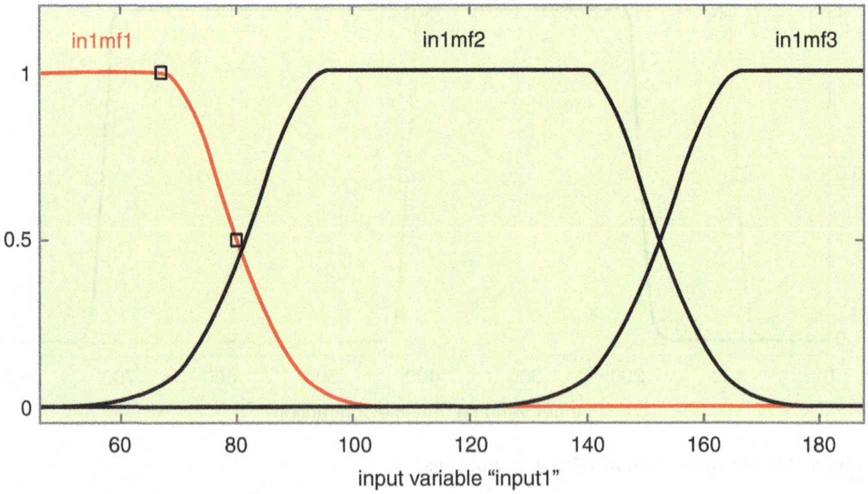

Fig. 9 Input1 membership functions

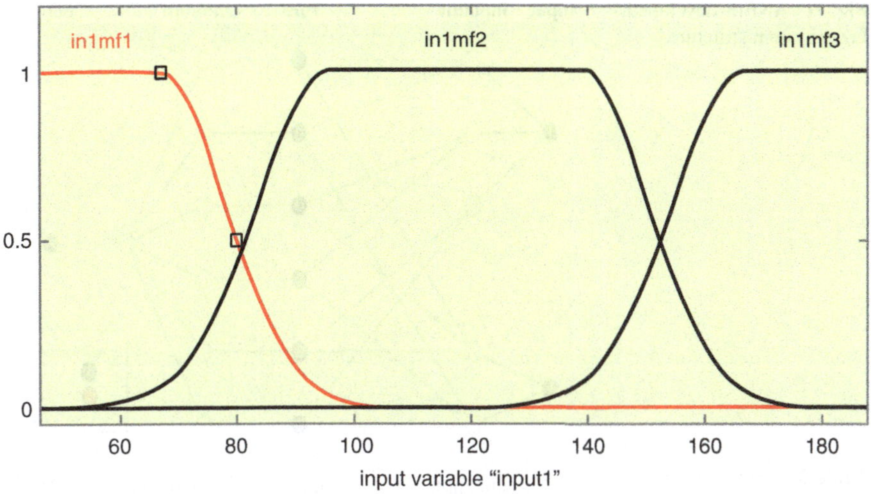

Fig. 10 Input2 membership functions

Table 2 ANFIS rules for the design

Rule	Heart rate	Muscle contraction	Classification
1	In1mf1	In2mf1	Out1mf1
2	In1mf2	In2mf2	Out1mf2
3	In1mf3	In2mf1	Out1mf3
4	In1mf1	In2mf2	Out1mf4
5	In1mf2	In2mf1	Out1mf5
6	In1mf3	In2mf2	Out1mf6

The ANFIS-developed fuzzy system structure in the form of neural network is shown in Fig. 11, having one collective set of artificial neurons. ANFIS have used the AND logic or intersecting membership function characteristics to develop decisions for classification. For all computational intelligence techniques presented in this study, same data sets were used: 21,400 samples for training and 200 samples for testing.

3 Classification Using Computational Intelligence

3.1 Artificial Neural Network

A total of 21,400 were experimentally gathered using MATLAB. Dual binary digit was implemented as output neuron establishment representation, in which "10" denotes data for having mental fatigue and "01" for not having mental fatigue. Samples of training data are presented in Table 3.

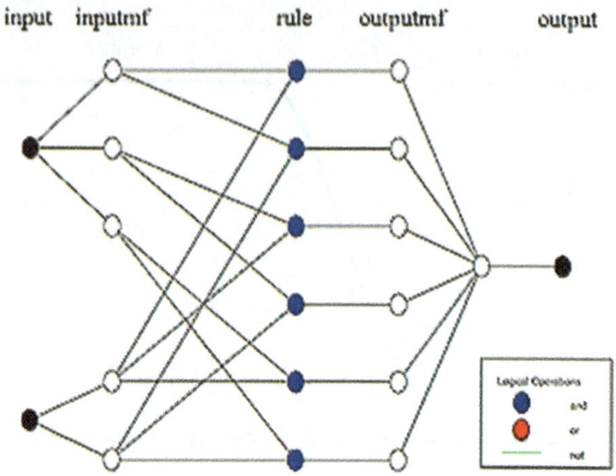

Fig. 11 ANFIS-developed fuzzy system structure

Table 3 Sample training data and results of backpropagation artificial neural network

Classification	Actual output		Artificial nodes (N)	
	N1	N2	N1	N2
With mental fatigue	1	1	0.9998	1.21E-05
Alert	0	0	1.06E-07	1.0000

Various network structures were developed in terms of number of hidden artificial neurons and layers to determine which configuration will yield the best system performance. Processing time, cross-entropy (CE) value, and regression coefficient (R) were considered determinants in finding out the best neural network.

By referring to Figs. 12, 13, and 14, it is noticeable that lowest CE is obtained from hidden node size of 10. The fastest learning time is obtained from hidden node size of 20. The regression coefficient obtained the highest value at hidden node size 140. Figure 12 shows that there is considerable increase in cross-entropy performance error for an increase in hidden node size. It means that cross-entropy diverts the predicted value from the actual value. Figures 13 and 14 depict that there is no particular increasing or decreasing linear pattern for learning time and regression coefficient in increasing hidden node size.

Figure 15 shows the training, validation, and test performance of the trained neural network. Its cross-entropy performance is 0.025389, indicating good performance of the network, wherein 0 is the ideal value. Twenty-seven epochs were used by the network to yield the best validation characteristic. During network training, the accuracy for training, validation, and test are 97.8%, 97.7%, and 97.9%, respectively. Hence, the overall training accuracy is 97.8%. After testing the developed neural network with different data sets from the training data set, the neural network classification accuracy yields 71.0%.

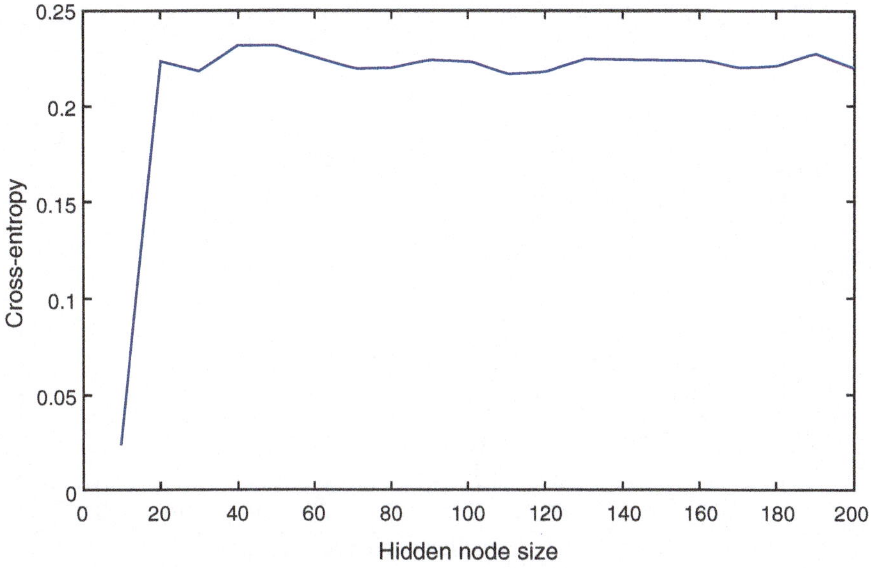

Fig. 12 Number of hidden nodes versus cross-entropy

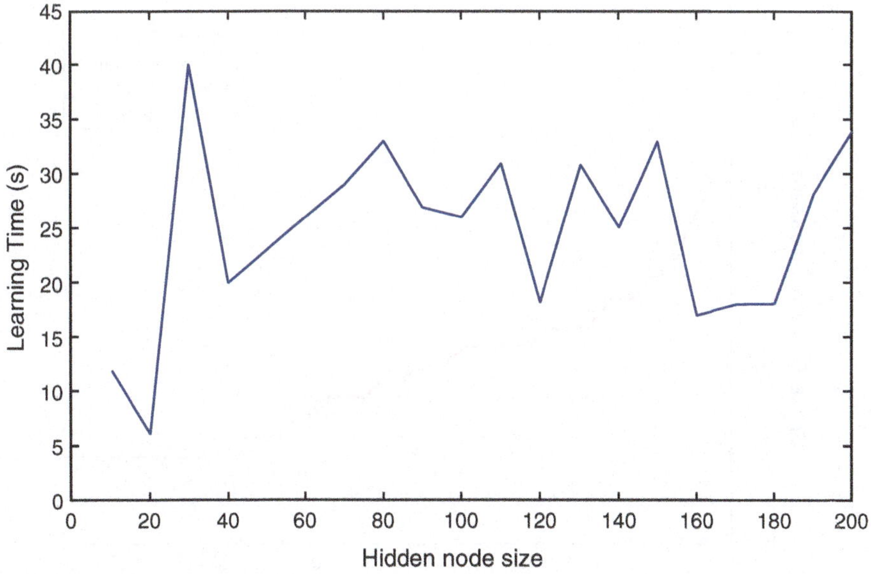

Fig. 13 Number of hidden nodes versus learning time(s)

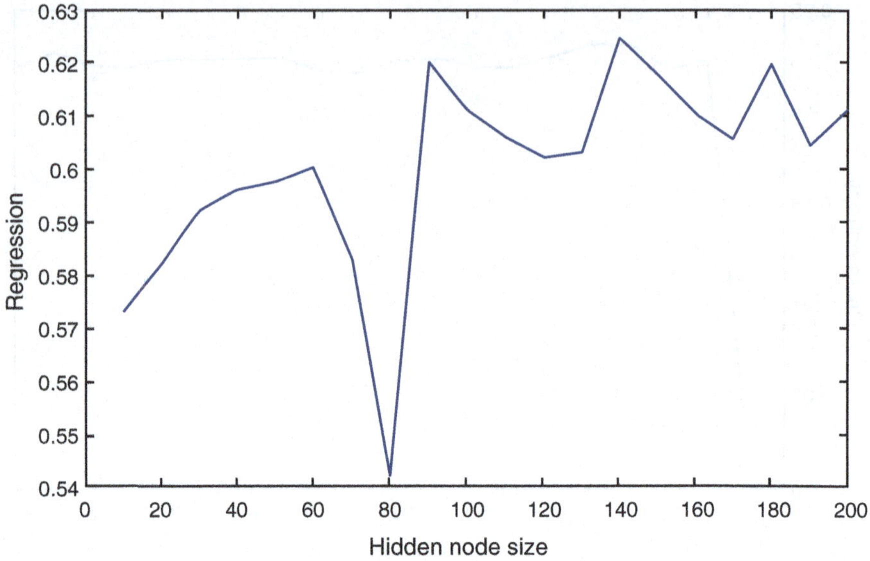

Fig. 14 Number of hidden nodes versus coefficient

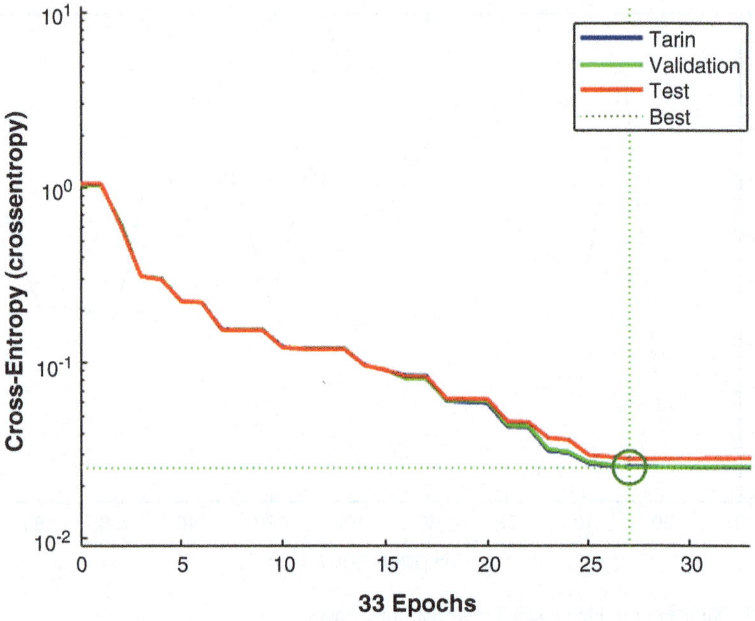

Fig. 15 Performance plot of the trained neural network with 10 hidden artificial neural neurons

3.2 Artificial Neural Network

Shown in Figs. 16, 17, and 18 are the curves obtained using the fuzzy logic system in MATLAB.

In fuzzy logic system, the activity level remains almost constant for heart rate starting at 80 beats per minute.

3.3 Adaptive Neuro-Fuzzy Inference System

Shown in Figs. 19, 20, and 21 are the curves obtained using ANFIS-based alertness and mental fatigue classification system.

By comparing the simulation outputs of fuzzy logic system and ANFIS, there is clear evidence that ANFIS provides a better classification mechanism. ANFIS characterized a smooth transition of brain activity level for every change in heart rate and muscle contraction. The activity level remains constant for heart rate starting at 80 beats per minute. The constant activity level is close to level 1, generalizing it as a person with good alertness level. Below 80 beats per minute, the network may consider it as transitioning period or with mental fatigue, as long as it is in tandem

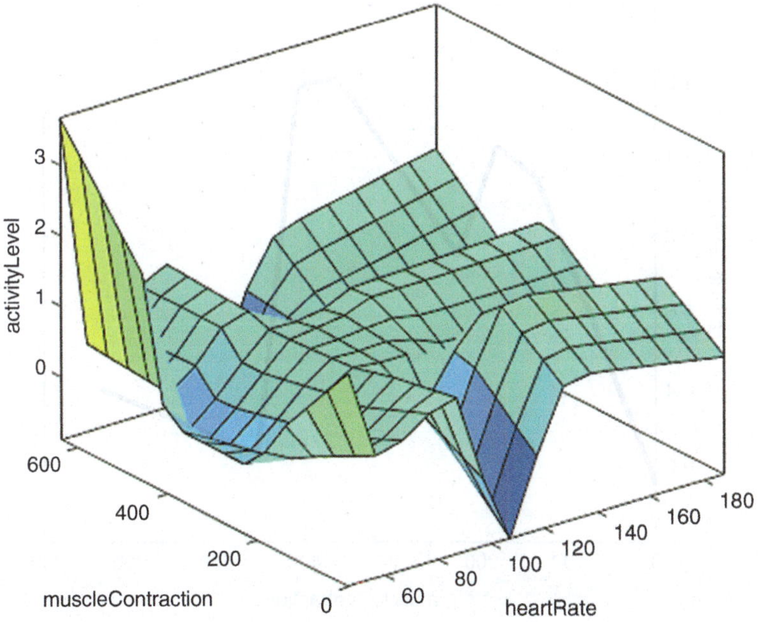

Fig. 16 Surface view using fuzzy logic system

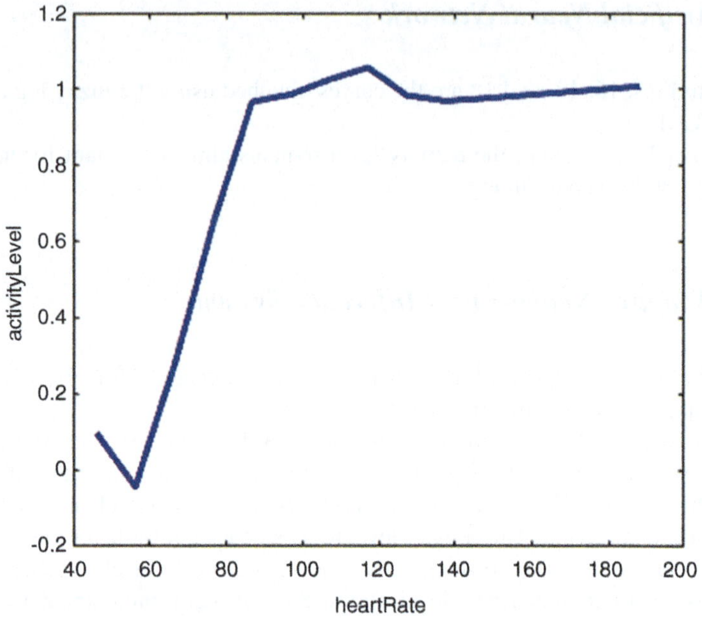

Fig. 17 Activity level with heart rate

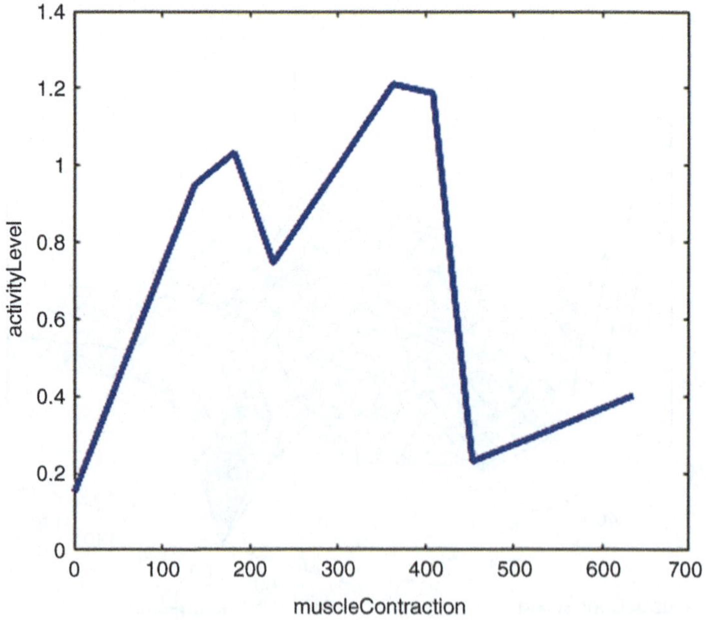

Fig. 18 Activity level with muscle contraction

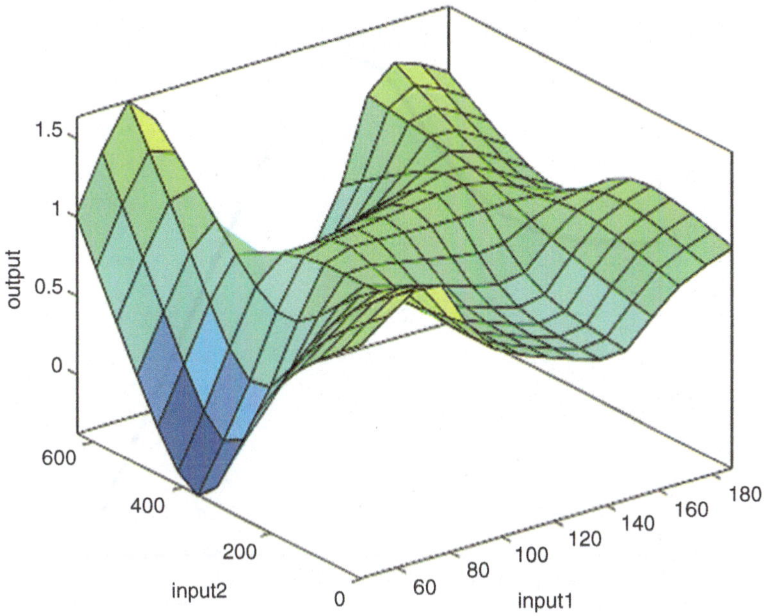

Fig. 19 Surface view using ANFIS

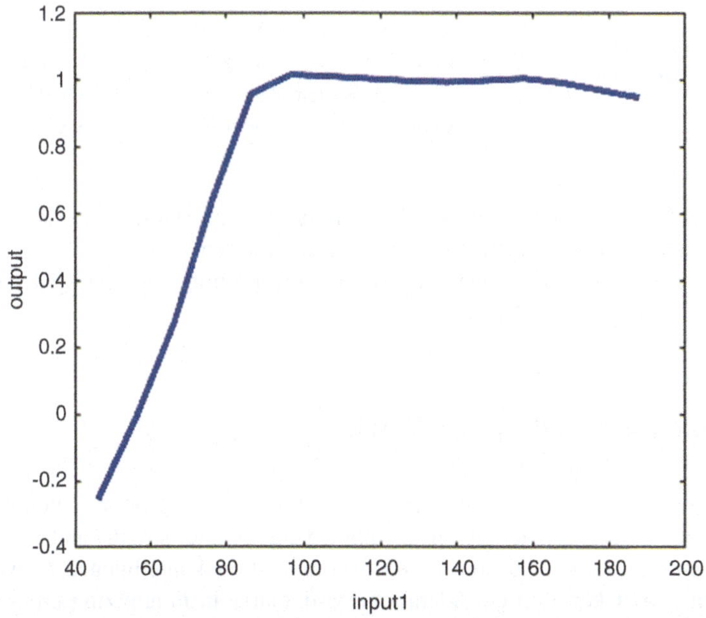

Fig. 20 Output with input1 (heart rate)

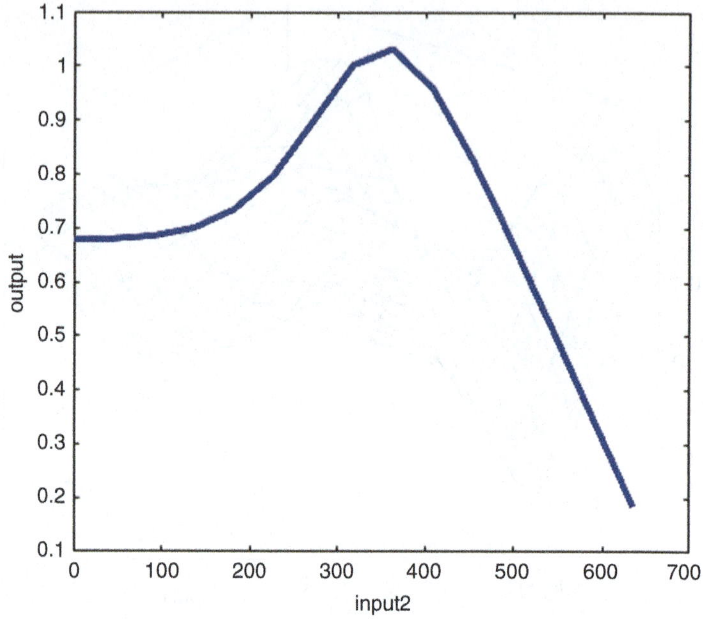

Fig. 21 Output with input2 (muscle contraction)

Table 4 Training and testing accuracy of developed computational intelligence techniques

Soft algorithm	Training accuracy	Testing accuracy
ANN	97.800%	71.000%
Fuzzy logic system	99.529%	99.553%
ANFIS	99.604%	99.556%

with muscle contraction criterion. The activity level constantly decreases in linear fashion, starting at 360 millivolts for muscle contraction.

Table 4 provides the numerical basis for the training and testing accuracy of ANN, fuzzy logic, and ANFIS.

4 Conclusion and Future Work

This study presented the suitability of applying computational intelligence in classifying human brain activity in a combined electrocardiography and electromyography system. Critical design of each computational intelligence network and algorithm was taken into consideration, particularly in minimizing cross-entropy error in training and testing. With the three soft algorithms, ANFIS provided the optimal accuracy both in training and testing of sample data, followed by fuzzy logic system and ANN. Adaptive neuro-fuzzy inference system yielded the superior classification mechanism for alertness and mental fatigue, as compared with artificial neural network and fuzzy logic system, with an accuracy of 99.556%. A power-efficient off-body area network was constructed using Bluetooth as the wireless medium.

Future works include using particle swarm optimization and simulated annealing technique to verify the most effective metaheuristic optimization method for this kind of data and the addition of skin conductivity sensor for more physiological data analysis and understanding.

Acknowledgments This project is supported by the Engineering Research and Development Technology program of the Department of Science and Technology and the Department of Electronics Engineering of the University of Perpetual Help System DALTA, Las Piñas Campus.

References

1. Chowdhury, A., Shankaran, R., Kavakli, M., & Haque, M. (2018). Sensor applications and physiological features in driver's drowsiness detection: a review. In *Proceedings of the IEEE JSEN, Sensors Journal.*
2. Salahuddin, L., Gi Jeong, M., & Kim, D. (2007). Dependence of heart rate variability on stress factors of stress response inventory. In *Proceedings of IEEE.*
3. Chai, R., Tran, Y., Craig, A., Ling, S., & Nguyen, H. (2014). Enhancing accuracy of mental fatigue classification using advanced computational intelligence in an electroencephalography system. In *Proceedings of IEEE.*
4. Zhang, C., Cong, F., & Wang, H. (2017). Driver fatigue analysis based on binary brain networks. In *Proceedings of the IEEE.*
5. Bergasa, L., Nuevo, J., Sotelo, M., Barea, R., & Lopez, M. (2006). Real-time system for monitoring driver vigilance. In *IEEE Transactions on Intelligent Transportation Systems.*
6. Wang, J., Xu, W., & Gong, Y. (2010). Real-time driving danger-level prediction. In *Engineering Applications of Artificial Intelligence.*
7. D'Orazio, T., Leo, M., Guaragnella, C., & Distante, A. (2007). A visual approach for driver inattention detection. In *Pattern Recognition.*
8. Grace, R. (2001). Drowsy driver monitor and warning system. In *Proceedings of the International Driving Symposium Human Factors Driver Assessment, Training and Vehicle Design.*
9. Borghini, G, Vecchiato, G., Toppi, J., Astolfi, L., Maglione, A., Isabella, A., Caltagirone, C., Kong, W., Wei, D., Zhou, Z., Polidori, L., Vitiello, S., & Babiloni, F. (2012). Assessment of mental fatigue during car driving by using high resolution EEG activity and neurophysiological indices. In *34th Annual International Conference of the IEEE EMBS.*
10. Bundele, M., & Banerjee, R. (2009). Detection of fatigue of vehicular driver using skin conductance and oximetry pulse: a neural network approach. In *Proceedings of 2nd WAS.*
11. Sangwa, A., & Bhattacharya, P. (2015). Wireless body sensor networks: A review. *International Journal of Hybrid Information Technology, 8*(9), 105–120.
12. Hanson, M., Powell, H. Jr., Barth, A., Ringgenberg, B., Calhoun, B., Aylor, J., & Lach, J. (2009). Body area sensor networks: challenges and opportunities. In *IEEE.*
13. Dangi, P., & Sangwan, K. (2017). Wireless body area sensor networks. In *International Conference on Computing and Communication Technologies for Smart Nation.*
14. Sinavandam, S., Sumathi, S., & Deep, S. (2007). *Introduction of fuzzy logic using matlab.* Berlin Heidelberg: Springer.
15. Sun, Z., Au, K., & Choi, T. (2007). A neuro-fuzzy inference system through integration of fuzzy logic and extreme machine learning. In *IEEE Transactions on Systems, Man and Cybernetics.*

Index

A. Beltran Jr. et al. (eds.), *World Congress on Engineering and Technology; Innovation and its Sustainability 2018*, EAI/Springer Innovations in Communication and Computing, https://doi.org/10.1007/978-3-030-20904-9

The manufacturer's authorised representative in the EU is Springer Nature Customer Service Centre GmbH, Europaplatz 3, 69115 Heidelberg, Germany. If you have any concerns regarding our products, please contact ProductSafety@springernature.com

Printed and bound by CPI Group (UK) Ltd, Croydon, CR0 4YY
30/09/2026
02225643-0020